# ISO 9000
# Quality Systems
# Development Handbook

*A Systems Engineering Approach*

# ISO 9000 Quality Systems Development Handbook

*A Systems Engineering Approach*

## David Hoyle

Butterworth-Heinemann Ltd
Linacre House, Jordan Hill, Oxford OX2 8DP
225 Wildwood Avenue, Woburn, MA 01801-2041
A division of Reed Educational and Professional Publishing Ltd

◊ A member of the Reed Elsevier plc group

OXFORD   BOSTON   JOHANNESBURG
MELBOURNE   NEW DELHI   SINGAPORE

First published 1998

© David Hoyle 1998

All rights reserved. No part of this publication
may be reproduced in any material form (including
photocopying or storing in any medium by electronic
means and whether or not transiently or incidentally
to some other use of this publication) without the
written permission of the copyright holder except
in accordance with the provisions of the Copyright,
Designs and Patents Act 1988 or under the terms of a
licence issued by the Copyright Licensing Agency Ltd,
90 Tottenham Court Road, London, England W1P 9HE.
Applications for the copyright holder's written permission
to reproduce any part of this publication should be addressed
to the publishers

**British Library Cataloguing in Publication Data**
A catalogue record for this book is available from the British Library

**Library of Congress Cataloguing in Publication Data**
A catalogue record for this book is available from the Library of Congress

ISBN 0 7506 2562 7

Typeset by Butford Technical Publishing, Bodenham, Hereford
Printed and bound in Great Britain

FOR EVERY TITLE THAT WE PUBLISH, BUTTERWORTH-HEINEMANN
WILL PAY FOR BTCV TO PLANT AND CARE FOR A TREE.

# Contents

| | | |
|---|---|---|
| **Preface** | | **ix** |
| | | |
| **1** | **Establishing the case** | **1** |
| | Alan's story | 1 |
| | The need for management systems | 12 |
| | The emergence of management systems | 15 |
| | Standards for management systems | 16 |
| | Quality systems and the small business | 17 |
| | Summary | 20 |
| | | |
| **2** | **Quality system principles** | **21** |
| | The established definitions | 21 |
| | Breakthrough and control | 22 |
| | The need for objective evidence | 25 |
| | Quality is a result of the business | 26 |
| | Customer needs and expectations | 26 |
| | Failure prevention system | 27 |
| | Causing success | 28 |
| | Using foresight | 29 |
| | Error detection and correction | 29 |
| | Quality is achieved through a chain of processes | 30 |
| | Behavioural system | 31 |
| | Technical or administrative system | 32 |
| | Quality systems or quality management systems | 33 |
| | Quality systems and Total Quality Management (TQM) | 34 |
| | Summary | 42 |

## 3  Preparing for project launch — 45

Project overview — 45
Exploration — 47
Project feasibility — 49
Appointing the Project Manager — 72
Common mistakes at the pre-launch stage — 74
Summary — 77
Pre-launch task list — 78
Pre-launch questionnaire — 79

## 4  Managing the project — 81

Project management strategies — 81
Project planning — 82
Organizing the project — 89
Quality information system — 96
Capturing existing practices — 98
Education and training — 99
Creating and sustaining the environment — 100
Organizing the assessment — 107
Common mistakes in project management — 109
Management task list — 111
Management questionnaire — 113

## 5  System design — 115

The nature of systems — 115
System requirements — 117
Quality system requirements — 118
System design strategies — 121
System development process — 126
Flowcharting conventions — 129
Business interfaces — 130
System interfaces — 132
The system model — 133
Function deployment — 141
Analysing business processes — 143
Deploying the requirements — 151
Process modification — 154
Using software tools — 157

| | | |
|---|---|---|
| | Deploying the product/service characteristics | 159 |
| | Identifying documentation needs | 162 |
| | Practical document identification | 178 |
| | Summary | 180 |
| | System design task list | 182 |
| | System design questionnaire | 183 |
| | Annex A: Documents required by ISO 9001 | 184 |

## 6 Documentation development — 187

| | | |
|---|---|---|
| | Document requirements | 187 |
| | Development planning | 188 |
| | Documentation methods | 189 |
| | Producing effective documentation | 190 |
| | Documentation tools | 193 |
| | Documentation standards | 193 |
| | Creating the quality policy | 204 |
| | Creating the Policy Manual | 206 |
| | Creating procedures | 212 |
| | Creating the organizational documentation | 223 |
| | Creating and using forms | 226 |
| | Summary | 228 |
| | Documentation task list | 229 |
| | Documentation questionnaire | 230 |
| | Annex A: Specimen procedure in flowchart form | 232 |
| | Annex B: Specimen documentation standard | 233 |
| | Annex C: Specimen role description for a process owner | 236 |

## 7 Document and data control — 239

| | | |
|---|---|---|
| | Document control strategies | 239 |
| | Database design | 240 |
| | Document control requirements | 244 |
| | Document development procedures | 246 |
| | Document change procedures | 250 |
| | Document accessibility | 257 |
| | Obsolete documents | 257 |
| | Control of external documents | 258 |
| | Controlling data | 259 |
| | Summary | 262 |
| | Document control task list | 263 |

| | |
|---|---|
| Document control questionnaire | 265 |
| Annex A: Specimen document development procedure | 266 |
| Annex B: Specimen document change procedure | 271 |
| Annex C: Specimen document change request | 277 |
| Annex D: Specimen document change notice | 278 |
| Annex E: Specimen external document control procedure | 279 |

## 8  The quality system management process  285

| | |
|---|---|
| The process model | 285 |
| System implementation | 285 |
| System implementation task list | 294 |
| System implementation questionnaire | 295 |
| System evaluation | 296 |
| System evaluation task list | 321 |
| System evaluation questionnaire | 322 |
| System maintenance | 324 |
| System maintenance task list | 335 |
| System maintenance questionnaire | 336 |
| System improvement | 337 |
| System improvement task list | 342 |
| System improvement questionnaire | 343 |
| System management organization | 344 |
| Summary | 349 |

## 9  ISO 9001 Exposition  351

**Glossary**  399

**Index**  407

# Preface

Many quality systems have been established by implementing the requirements of ISO 9000 but they have frequently resulted in task-based systems of documentation rather than documented management systems. The 1990s has seen a rise in popularity of techniques such as TQM, Business Process Re-engineering, and Process Management, thus causing a rethink of ISO 9000 based quality systems. This book shows how you can use systems engineering principles combined with process flowcharting techniques to design and develop an integrated quality system that reflects how you run your business. Details are provided on how to produce and manage systems using both paper and electronic-based documentation, with guidance provided on designing and selecting various software packages.

This is the third book in a trilogy commencing with the *ISO 9000 Quality Systems Handbook*, published in 1993. The first book primarily explored each individual requirement of ISO 9001 and was intended for implementers. Part 1 covered general quality concepts, an anatomy of the standard and brief details of system development and assessment. The elements of system assessment were further developed in the second book, *ISO 9000 Quality System Assessment Handbook*, published in 1996 and the elements of system development are now covered in greater detail in this third book. I first started work on the trilogy in 1992 after completing 22 years in the quality profession, 20 of which were in industry. I had worked with standards similar to ISO 9000 for most of this time and started developing quality systems to meet ISO 9000 in 1987. I hadn't at that time developed a methodology – that was not to come about until 1990, when I was working with the UK Civil Aviation Authority as a management consultant. By 1994, when I intended to launch this book to complement my first book, my development methodology was still not mature enough and it was to be another four years before I was able to start on the third and final book in the trilogy. Over the same period, others have developed and used Business Process Mapping, Deployment Flowcharting and other techniques. I was unaware of these techniques back in 1990 and at that time was trying to analyse operational tasks that did not appear to follow a logical order. All the time I had this notion that we had to develop a system, not a collection of parts, and this has always been my driving force. I had spent much of my time in industry, in the project management and system engineering side of the business and it therefore came naturally to me to apply system engineering and project management techniques to quality system development. It was not until I came out of the CAA in 1992 to run auditor training classes and conduct audits in other companies that I realized how uniform ISO 9000 quality systems had become.

In the ten years since the publication of ISO 9000, over 130,000 organizations throughout the world have gained registration[1]. One could therefore say that the timing of this book on quality system development is somewhat off beam! Many books have been written on ISO 9000 so what else can one say? It is my belief that few of these registrations were of effective systems. The documentation may well meet the requirements of the standard and may also be followed, but this does not necessarily mean that a system is in place.

Around the world, there are many types of quality system but one dominates the field – the ISO 9000 element-based quality system. Where it originated is uncertain, but it is advocated with monotonous regularity. This approach facilitates the comparisons of various systems but it defeats the objective of providing systems that enable organizations to design, develop, produce and deliver products and services that meet customer needs and expectations. Quality systems should be developed to serve the needs of the business – not the auditors. Many of the software packages available for quality system developers have built-in prescriptions and templates for documentation which may not be appropriate to your organization and may force you into the ISO element approach.

In the last ten years managers have begun to realize that it is the process that creates the output, not the task, and an undue concentration on tasks rather than processes does not in fact bring about the benefits they set out to achieve. Drucker says[2] that: 'There is more to increasing productivity in knowledge work and service work than defining the task, concentrating on the task and defining performance. We do not yet know how to analyse the process in jobs in which performance predominantly means quality. We need to ask instead, "What works?" For jobs in which performance means both quality and quantity, we need to do both: ask what works and analyse the process step by step and operation by operation.'

Michael Hammer on the same subject says[3]: 'It is no longer necessary or desirable for companies to organize their work around Adam Smith's division of labour. Task-oriented jobs in today's world of customers, competition and change are obsolete. Instead, companies must organize work around process.' Some say good managers can manage well without really knowing much about what they are managing. It is the skill that

---

[1] *Mobil Survey of ISO 9000 and Environmental Certificates Awarded Worldwide* (August 1996).
[2] From *Managing the Future, The 1990s and Beyond*, Peter F. Drucker (1992).
[3] From *Reengineering the Corporation*, Michael Hammer and James Champy (1993).

counts. However, as David Packard remarks[4]: 'I don't argue that the job can't be done that way, but I do argue strongly that the best job can be done when the manager has a genuine and thorough understanding of the work. I don't see how managers can even understand what standards to observe, what performance to require, and how to measure results unless they understand in some detail the specific nature of the work they are trying to supervise.' And I would add that an understanding of the nature of the work is brought about by a knowledge of the various processes that produce the business outputs.

ISO 9000 addresses a series of topics that companies have followed too closely and not observed the intent of the series of standards. What ISO 9000 has done is to provide a prescription for success that companies can follow. The net result has been a collection of documents that describe tasks rather than processes. The standards do not in themselves give companies a method that they can follow. Without a method, companies have only documented what is needed in order to comply rather than describe the processes that produce the business outputs. Hence there is no connectivity between the tasks. All tasks are performed within a process to convert given inputs into outputs. But as Gregory Watson explains[5]: 'In order to re-engineer a process, a thorough understanding of the business is not required – only of a business process. However, in order to engineer a business system, the architect of change must consider all of the elements of the business and not be restricted to a singular process. This is why the engineering of an entire business system requires a cross-functional approach and also why the participation of the three major cross-functional support groups must be co-ordinated: quality, human resources and information systems.' Whilst this book does not cover business systems per se, it is perceived that a quality system cannot and should not exist in isolation of other business systems. Quality is a result of the business and hence everyone in the business contributes towards that result. Hence the quality system must be integrated with the other systems to become one integrated management system.

Why a systems engineering approach rather than a process management approach? As illustrated by Gregory Watson, concentrating on processes only requires an understanding of the processes and not the business. We need to understand how the business functions, what processes are required to deliver the outputs and how they work together, as a malfunction in one process can adversely affect the performance of interacting processes. Although business systems comprise processes, we need to apply systems engineering techniques to design a coherent system that fulfils the needs of the business. The development of a quality system or any type of business system requires the same project management techniques as would be used if designing a computer system. We need to manage cross-functional teams in order to produce a composite system, not a 'collection of parts'. We need a system requirement and to conduct design reviews to verify this requirement is achieved. We need to control the configuration of the quality

---

[4] From *The HP Way*, David Packard (Harper Collins, 1995).
[5] From *Business Systems Engineering*, Gregory H. Watson (1994).

system as change in the design of one process may well affect others, as stated above. We need to evaluate and qualify the system before we launch it and there are many other system engineering techniques that we can apply that will ensure we introduce a well-engineered quality system into the business.

For those who have already installed an ISO 9000 based quality system and are looking at ways to improve its effectiveness, or extend the system to encompass environmental, safety and other elements, the approach advocated in this book provides the techniques to re-engineer these quality systems and develop an integrated management system.

It has been a long-held belief of mine that well-engineered quality systems will improve business performance and drive changes in culture as well as in products and processes. If you embrace the intent of ISO 9000 rather than restrict yourselves to the requirements, change of any nature can be brought about by a well-engineered quality system.

The book addresses the mechanics of quality system design through to system implementation and subsequent improvement. It includes many system models, process charts and procedures to illustrate the design principles involved.

The first chapter establishes the case for a quality system through the experiences of a sales executive and leads on to show how management systems emerge and the management system standards evolved. Comparisons are made between ISO 9000 and TQM showing that most of the TQM principles can be drawn out of ISO 9000 if one follows the intent of the standard.

The second chapter covers the basic principles underlying quality system development and is followed by two chapters covering the preparation for and management of quality system development.

Chapters 5 and 6 address system design and documentation and Chapter 7 the techniques used to control the documentation and data produced and utilized by the system. Specimen procedures and forms are provided to illustrate various styles. Over the last few years many software packages have been developed to assist quality system designers and guidance is provided in the design and use of such tools.

The management of the quality system has often been perceived as a documentation task. In Chapter 8, all the processes that serve quality system management are brought together to show how they interact to develop, evaluate, maintain and improve the system. Process flowcharts for each of the business and work processes are provided with supporting descriptions pointing to the key aspects that need to be addressed in developing these processes.

Most chapters include task lists and questionnaires to assist system designers and project managers in the system development project.

Chapter 9 breaks the sequence and style of the book. It contains an Exposition covering the requirements of ISO 9001. Many quality manuals respond to the requirements of the standard with less description than you will find in this chapter. The Exposition is not intended as a replacement for a quality manual, but a supplement. The responses can be used to compile a manual that mirrors the way you do business.

There is also a glossary of terms that are used in the book.

This is not a book about ISO 9000, or about TQM, or Business Process Re-engineering (BPR), even though one will find these topics covered. This is a book about the design and development of management systems, particularly quality systems. That in itself is not new. Many approach the subject of ISO 9000 in terms of ISO 9000 being implemented rather than a system being designed that meets ISO 9000 requirements. There is a fundamental difference between these two approaches on which I will elaborate.

A common theme throughout the book is that a system should not only prevent problems but also cause the right things to happen and I believe that an effective way to develop a quality system that fulfils this purpose is to use a systems engineering approach.

David Hoyle
Monmouth, April 1998

# Acknowledgements

This book is the product of many years working in quality management for large and small companies. The ideas started to come together whilst I was with Ferranti Computer Systems, having navigated through six reorganizations in as many years and several Ministry of Defence Contractor Assessments. This experience impressed upon me two principles regarding quality system design: the need to make management systems as immune to organizational change as possible and the idea that a management system is not a collection of parts but an interconnection of processes. Although the company no longer exists in its previous form, these experiences, whilst stressful, were truly enlightening and so deserve acknowledgement.

My time with the Civil Aviation Authority caused me to refine my thoughts on quality system design and the encouragement and support I had from Neil Menzies of the Scottish Air Traffic Control Centre and Ray Tindell of the London Air Traffic Control Centre. I am also indebted to Neville-Clarke International who provided me with countless opportunities world-wide, either through auditor training or consultancy assignments to see first-hand how companies had implemented ISO 9000. In particular Bob Birtwhistle of Neville-Clarke for the many enlightening discussions we had on the Business Process oriented training material we developed and delivered.

Three clients in particular have helped enormously to refine my thought processes and try out new ideas. The first was Philip and Son, a small ship repair yard and my thanks to Ali Khatir who was thrilled to try something new. Secondly to Ray Mann and staff of the Mann Organization whose operation appeared so complex that it lent itself to the integrated systems approach. A special acknowledgement to Bob Antonetti, Andrea Green and Melissa Matthews of the New York and New England Telephone Exchange (NYNEX). After explaining the principles to them, they were in no doubt that the systems engineering approach was the most effective way to develop their quality system as a system arranged around the elements of ISO 9002 had no meaning to them. They supported, promoted and embraced the approach to a degree that when the formal assessment came they sailed through without any nonconformities. My appreciation also to the Mann Organization and to NYNEX for giving their permission for me to reproduce some of the flowcharts and procedures we developed, although no doubt they have changed a little since then!

I am grateful to the Institute of Quality Assurance for providing a continual source of reference material through *Quality World* and for publishing several papers I submitted on quality system development in the last two years. The IQA have also given me the opportunity to develop and deliver training courses on the techniques described in the book. The ASQ through *Quality Progress* have also been an excellent source of technical papers that were helpful in clarifying ideas and beliefs.

Finally, I express my gratitude to my wife Angela who made a significant contribution to the content of this book and who continues to provide support and encouragement.

---

Other books by the same author:

| | |
|---|---|
| *ISO 9000 Quality Systems Handbook* | ISBN 0-7506-2130-3 |
| *QS-9000 Quality Systems Handbook* | ISBN 0-7506-9861-6 |
| *ISO 9000 Quality System Assessment Handbook* | ISBN 0-7506-2563-5 |

---

Readers are invited to contact the author with any questions, comments and constructive criticism:

*by e-mail at:* hoyle@celtic.co.uk

# Chapter 1

# Establishing the case

## Alan's story

What makes a satisfied customer? It's a difficult question to answer as each customer has different wants and needs.

Alan was a sales executive and his job was to meet his sales targets for the period. He had done well these last few months as he had passed his target with no trouble. He felt this was due to his company's excellent product, and so it was. The new model was indeed a breakthrough in technology and surpassed anything the competition could rival. It had been in production now for over a year and was selling well.

Then one day a customer called to say his machine was faulty. It was only three months old and was therefore covered by the warranty. Alan told the customer to return it. 'We will repair the product in five days or replace it,' he told the customer (this was company policy). Alan had always put the customer first. He was not the kind of sales executive who would try to win a sale by cheating the customer. He believed in the win-win situation and would never fool a customer into believing that he was getting a bargain when he wasn't.

Five days later the customer called again to say that a replacement had not been supplied. Alan called the Servicing Department and was informed that there had been a backlog of repairs and it would be another two days before they could get around to that particular machine. He notified the customer, who was not at all pleased, as no replacement was offered as promised. Alan apologized and said he would get back shortly. He began to think about the company policy of 'We will repair the product in five days or replace it'. He knew of no mechanism in place that would guarantee a replacement within the five day period. Meanwhile, another of Alan's customers had a faulty machine so he started an investigation to find out whether the two machines had a common fault. He found that in both cases a fault was discovered in the same component. Further investigation established that a batch of machines had been assembled by a new employee who had misunderstood the part numbers and consequently fitted the wrong part. In fact, it was the right type of component but did not have the latest modifications. In discussing the problem with the Production Manager, they agreed that all new employees would be given training in the company's numbering system.

The second customer had bought the machine over twelve months previously and it had had the older part fitted – it was one of the first production machines. However, that machine was no longer covered by the warranty so Alan had to explain that there would be a repair charge. The customer was not pleased as the warranty period had only just expired, and enquired as to the nature of the fault. Alan explained that the model had been modified (on reflection he thought he shouldn't have told the truth but that was not his nature). The customer naturally was upset and demanded a no-charge repair since the fault was not due to fair wear and tear. Alan checked with his superior and was told that a charge would have to be made since they could not have one law for one customer and another for the next. Needless to say the customer was very dissatisfied and threatened to take his business elsewhere next time.

Alan's first customer was on the phone again. Two days had elapsed and no replacement machine had been delivered. Alan again investigated the problem and found that several people in the Servicing Department had gone sick and the backlog was increasing.

'Why can't we use the production line technicians?' Alan asked Bill, the Production Manager.

'Well, we could,' replied Bill, 'but they are working flat out to meet their quota.'

'Why can't we take in contractors to move the backlog?' enquired Alan.

'You'll have to take that up with Tony, the Servicing Manager,' said Bill. 'I am only responsible for production, not servicing.'

Alan went into the Servicing Department and put the problem to Tony.

'We can't move production staff into servicing,' said Tony, 'because they won't understand the way we do things around here.'

'But haven't we got written procedures or something that they can follow?'

'Hell no!' said Tony. 'I haven't time to write things down and anyway it would always keep changing. You know how lean we are. The slightest change and everything is affected.'

Alan began to reflect on his week so far. There had not been so many problems in a single week before. There were the faulty machines, although they had tracked down the fault and hopefully taken adequate corrective action by training all new employees. Then there was the resource problem in the Servicing Department. They couldn't solve that until either they had more technicians or could use contract labour. If they had documented procedures, Alan thought, new staff and contractors could quickly learn their methods, so he undertook to speak to the Chief Executive Officer (CEO) about this. The CEO told Alan that his job was to satisfy customers and not to worry about operations, as others would write procedures if they believed they needed them. He also made it clear that Alan was not to interfere in the work of other departments.

'Satisfy customers,' Alan thought. 'Well, we haven't done very well this week.' He started to list the incidents.

- Delivered several faulty machines.
- Didn't meet the response time promised.
- Didn't provide the replacement we were committed to.
- Made a customer pay for something that was our fault.
- Inadequate resources in the servicing department.
- Inadequate training of new employees.
- No written procedures.

Alan began to think about customer satisfaction. What satisfies the customer? He came to the conclusion that it is more than supplying product that is ordered. 'The products have to be reliable and if they do happen to fail then the customer expects this to be put right at our cost if its our fault. But not only that, the customer also expects a prompt service and, if we make promises, we should keep them.'

Later the following week Alan checked the problems previously encountered to see if the agreed actions had been taken. He was pleasantly surprised to find that both the faulty machines had been repaired and dispatched. He called the first customer and was told that the machine was delivered but not in the five days promised. However, the machine worked so the customer was relatively satisfied. In the second case the customer was livid. Not only was a charge made for labour but for a new part as well. Alan said he would withdraw the invoice and passed a memo to Accounts to execute it.

A few weeks later Alan received yet another complaint from a customer with a faulty machine. Alan's chance meeting with Bill in the canteen prompted him to ask whether the training of the new employee had been initiated. Bill was puzzled by Alan's question since, as far as he could recall, the training was scheduled for only one person. Alan was puzzled by Bill's response. Surely Bill had installed a training plan that would be implemented on every new hire. As it turned out, no such plan was implemented because, at the time, Bill had only one new employee in the Production Department. However, since then Bill had replaced one of his staff, as a sick employee had not returned to work. When Alan looked into the problem he found that the last machine that was returned had been assembled not by a new employee but by a worker who had transferred from the Servicing Department. On further investigation it transpired that an operator had again misunderstood the company's part numbering system. Alan burst into Bill's office and demanded an explanation. Bill told Alan that the numbering system was inherently troublesome as it was not logical.

'Anyway, I expected one of Tony's servicing staff to understand the numbering system,' replied Bill to Alan's outburst.

Alan asked why a new system had not been developed and was told to his dismay that there were more important things for the development engineers to do. 'But it's losing customers,' said Alan.

'Well you had better go and tell the Chief Designer because every time I mention it I am criticized for employing idiots,' said Bill.

A month later the CEO announced that the repair group in the Servicing Department was transferring into production in order to overcome the resource problems and provide greater flexibility. On hearing the news, Alan went to see Bill and asked him what he intended to do about training the new staff. Bill told Alan that he hadn't got the time and anyway, as he had said previously, 'These people know the job!'

'I wouldn't have agreed to the transfer if I thought I had to train them,' said Bill. Alan could see he was not going to get a commitment from Bill so left his office.

Alan began to think about how the company could ensure these failures didn't recur and thought out two solutions. One way was to train people in the company's methods but even this was unreliable since it relied on the methods being practical in the first place. What was really needed was a change in the numbering system but when Alan approached Andrew, the Chief Designer, he was informed that it would affect every product, would be too costly and no allowance had been made for it in that year's budget. Alan began to think that no one seemed to be interested in preventing errors and came to the conclusion that each person was only interested in his/her own department's performance.

Several weeks went by and more complaints came in, although of all the machines sold to date they were but a small percentage. However, some of the machines were from the initial batch with the faulty component. Alan began to think that the company should have recalled these machines and fixed the fault but the Chief Designer insisted that the fault was not predictable and only occurred under certain operating conditions. It was therefore more economical to replace the component when a customer noticed the fault rather than replace every component regardless of a fault being detected. Alan felt this attitude did not demonstrate a commitment to quality but chose not to pursue the matter as his annual appraisal was imminent and he didn't want to rock the boat at this particular time!

Then, in one week, three customers complained about faulty machines. Alan investigated the faults and found that all three machines had been assembled by operators who had transferred from the Servicing Department a month previously. Alan went to speak to Bill. On entering Bill's office he was surprised to find that Bill had moved to the sister plant and Tony had taken over Production. Alan explained to Tony the problems they had encountered the previous month. Bill had not passed on these experiences and Tony knew of no training plans or problems with the numbering system. In fact, he had reorganized production to make it more efficient and everyone was now working to Servicing Department methods.

Alan left Tony's office in despair. He suddenly remembered the overcharged customer and called Accounts. They didn't have any record of a cancelled invoice. How could he be responsible for customer satisfaction when the means of achieving it were so unreliable? He began once more to reflect on the past and added a further three notes to his list of incidents.

- More concerned with departmental budgets than eliminating failure.
- Methods change when managers change.
- No feedback on actions.

After work one night, Alan went to a lecture on quality hosted by the local Chamber of Commerce. The speaker was a CEO of a company that had recently been awarded an ISO 9000 certificate. He spoke of the benefits that such an award brings in terms of increased marketability and, although it had cost quite a lot, he remarked upon how easy it was to gain certification. All they had to do was document what they do, do what they documented and prove it. Alan thought he may be on to something here and managed to corner the speaker to talk with when his talk ended.

'Yeah! That's right,' said the speaker. 'It's easy, although you have to carry out audits periodically and management have to review the system but that's no hardship. We do that after the board meeting at the end of each quarter.

'What is ISO 9000?' asked Alan.

'It's an international standard,' said the speaker. 'All companies are going for it. You can't afford not to in this competitive age.' The speaker gave Alan his business card and suggested he visit the factory to see for himself.

Although Alan had other work to do, he felt he ought to investigate this 'ISO 9000 thing' and so he called the local Chamber of Commerce and asked for details. They said they would send Alan an information pack. When it arrived Alan studied it in depth and, being a cautious person, he noticed that there were several standards referenced. After writing off for several of the documents he began to wonder how documenting what you do could possibly improve quality when they were not doing the right things in the first place. He remembered the speaker at the lecture and retrieved his business card.

John Penfold was a busy executive with little time to spare but welcomed Alan's call, inviting him to the factory the next day.

'Unfortunately I will be out of town,' said John. 'But I will inform my QMR and she will show you around. Her name is Claire Hughes. She's on the ball and knows all about ISO 9000.'

Alan arrived at the plant the following morning and met Claire. Enthusiastic to show off the new award, Claire showed Alan the certificate that was in pride of place in the entrance hall. Alan noticed that the certificate was named ISO 9002 not ISO 9001 and queried this with Claire. Claire explained that whilst they do design their own products they don't design products for specific customers and therefore ISO 9002 was the appropriate assessment standard.

'Assessment standard?' queried Alan.

'Yes,' said Claire. She went on to explain that there are three assessment standards, ISO 9001, ISO 9002 and ISO 9003, and they are used for contractual and assessment purposes.

'What about the other standards such as ISO 9004?' asked Alan.

'Oh! I see you know your standards,' said Claire. 'We didn't use ISO 9004. Its only a guide and not used in the assessment.'

Claire and Alan then proceeded to her office, where she showed him the manuals and procedures. Alan was impressed.

'One manual and twenty procedures,' said Alan. 'That's not very many documents to produce.'

'We had some help from a local consultant who provided some sample documents but it is largely all our own work,' said Claire.

'Your CEO said that all you had to do was to document what you do and do what you document. Is that what these documents represent?' asked Alan.

'Well, not quite,' said Claire. 'John has a habit of oversimplifying the task but we did limit ourselves to the documents that are required by the standard.'

'One thing I recall from reading the standard,' said Alan, 'is that it focuses on the prevention of nonconformities.'

'Yes, that's right,' said Claire 'and we do have a preventive action procedure – look I'll show you.'

Alan examined the procedure and was delighted to see references to customer complaints but there was no reference to training. 'Why does this procedure not cover staff training?' he asked.

'We have another procedure for that,' explained Claire.

'I don't understand, isn't training a preventive action?' queried Alan.

'Well, I suppose it is,' replied Claire. 'But the standard requires a separate procedure – look I'll show you.' Claire began to point out the requirements for a procedure to identify training needs.

'I see,' said Alan. 'A procedure for training and one for preventive action that covers other types of preventive action. How about design changes?' asked Alan.

'Oh, we don't cover design in our system,' explained Claire.

'Yes, I remember you telling me but I don't understand why you omit design changes when they are one of the most significant forms of preventive action.'

'Yes, I know,' said Claire, 'but our CEO wanted to do only the minimum to get the certificate.'

'You mean you can get the certificate without bringing design changes into your system simply because you want to omit it!'

'Yes, in fact it was the Registrar that told us about it,' replied Claire.

'I thought the Registrar was the body that conducted the assessment,' Alan remarked.

'Yes, it is,' said Claire, 'but they agreed the scope of registration with us and told us we could define our own scope – so we did.'

'Mm' thought Alan, 'I'm not so sure of the value of ISO 9000 now. Tell me Claire, I noticed in ISO 9002 that the system has to ensure that you meet specified requirements.'

'Yes, that's right,' said Claire.

'How many customer complaints do you have?' enquired Alan. 'Approximately, you understand, I don't want you to breach company confidentiality.'

'Oh! That's no problem. Every company has complaints,' said Claire. 'We had about 20 a month when we started but now it's dropped to 15 as a result of our ISO 9000 programme,' continued Claire, 'and that's in only 6 months. Our CEO is very pleased, as I'm sure he told you. The registrar was also pleased and complimented us on our achievement.'

'I thought that a quality system would enable us to eliminate customer complaints entirely,' Alan remarked, 'and it surprises me that you actually gained registration simply by showing that the level had declined. It does seem rather easy to obtain this certificate.'

Alan looked through some of the other procedures and noticed that they all had a title that corresponded with the section headings of the standard. 'Why is this?' enquired Alan. 'Don't you have procedures that tell people how to do their job?'

'Oh yes,' replied Claire. 'We have work instructions as well.' Claire proceeded to lift down from the shelf the Production Department Manual. 'We have others,' continued Claire, 'in fact one for each department.'

Alan skimmed through the manuals of procedures and noticed that they didn't follow the headings in the standard. 'Tell me,' said Alan, 'do you have any procedures that are common to all departments?'

'We have a few,' said Claire, 'but most departments have just written what they do and we found that several wanted to do things slightly differently. Take document control for example – we have a common procedure for the quality system procedures but each department has its own for controlling its own documents.'

'I don't understand,' said Alan. 'Wouldn't it be more effective if you had just the one system?'

'You can't do that around here,' said Claire. 'Every manager wants to have control over his or her own operations. Managers won't work to another's procedures. They say it reduces their flexibility.'

'What happens when some information passes from one department to another. Is it conveyed in a common format?' asked Alan.

*Establishing the case* 9

'Sometimes,' said Claire. 'We have a common Nonconformity Report we call an NCR. Although Inspection raise the NCR, it is passed to the department responsible for the error and they fill out the details and pass it back to Inspection.'

'What happens if the problem arises in Servicing and not Inspection, do they use the same form?' asked Alan.

'No,' replied Claire. 'Our Servicing Department needed to record repair times so they use a different form.'

'That's odd,' remarked Alan. 'Why are repair times of importance only to the Servicing Department – surely Production do repairs before shipment?'

'Yes they do,' replied Claire, 'but because the customer is not paying, they are not logged.'

After some further discussion about the elements of the quality system, Alan bid Claire farewell and thanked her for a most enlightening visit.

Back in the office, Alan mused over his discoveries. He took out the standards from his desk and read them again and again but was confused by the terminology and apparent ambiguities. 'We have to establish a documented quality system,' said Alan to himself. 'Prepare procedures and then implement them. That's what the standard states but it isn't enough. I need more information.' So Alan went to the library and borrowed some books on the subject.

In the evenings Alan read each of the books, made notes and frequently consulted the standards to confirm what he found. It seemed that it was more or less as Claire explained. Document what you do and do what you document and prove it. Twenty procedures or so and a manual that basically repeats what's in the standard. If its so simple why are there so many consultants offering their services to assist companies through the certification process? He then remembered the problems he had encountered at work and could not understand how such a system could resolve such problems. He was determined not to approach the CEO without a viable plan, so he continued his search for the solution.

When he returned to work he noticed that Bill had been into the plant on a flying visit and left a newspaper cutting on his desk. Alan studied the cutting. It was about the failure of ISO 9000 to bring about the promised benefits. Alan read that companies were claiming that their investment had not paid off. There was too much documentation and customers were complaining that all it did was to give companies a licence to deliver shoddy goods consistently. This last report reminded Alan of Claire's record of customer complaints. If some companies were supplying goods to their own requirements rather than meeting customer expectations then it was not surprising that customers were complaining.

'This can't be right!' thought Alan. 'How can all these companies be missing the point? It is clearly stated in ISO 9001 that it aims to achieve customer satisfaction through the prevention of nonconformity. Regardless of the specific requirements the aim is quite clear – prevention.' (Not detection and correction, as seems to be practised.) 'They have to establish a system'. There Alan paused. 'A system,' he thought. 'Where have I seen that before?'

Alan searched through his files for the standard on terminology. 'A system – where does it state what a system is?' he asked himself. 'It is not defined in ISO 9001, that's for sure. All the standard says is that suppliers have to establish a system and then goes on to define a series of requirements. But requirements are not systems,' Alan thought. Then he remembered his days at university where he studied Engineering.

In his notes, which he had kept in the attic, Alan found what he was looking for: *System Design Notes*. He read over what he had written many years before. '*A system is an ordered set of ideas or an interconnection of processes that achieve a given objective.*' 'That's what I want!' Alan exclaimed. 'Now all I need is a link into ISO 9000.'

Remembering a standard he had scanned previously he withdrew ISO 8402 from his files and searched for the term *quality system*. At last he found it and read '... *the organization, processes, procedures and resources for implementing quality management.*'

The standard did not mention interconnection of processes but there was some mention of it he was sure. He spent hours pouring over the standards until he discovered ISO 9004-4 on quality improvement. Here he found mention of processes in connection with quality improvement. '*Quality improvement,*' he read, '*is achieved by improving processes. Every activity or item of work in an organization comprises one or more processes.*' Further on he read of an effective way of organizing quality improvement that identifies opportunities both vertically and horizontally in the processes that flow across organizational boundaries. Now he had the link. 'The pursuit of ISO 9000 must be a quality improvement – if not, why go for it?' he thought. 'And if it is a quality improvement, then it should be treated as such and be organized as explained in ISO 9004-4, around processes that flow across organizational boundaries.'

A system wasn't a collection of procedures but the interconnection of processes. ISO 8402 mentioned the organization. 'Yes, so it ought to do,' thought Alan, 'after the problems we had when the production manager was changed.' Then procedures were mentioned and Alan remembered well the need for procedures so that managers could rely on staff doing things the same way. In spotting the word 'resources', Alan remembered the staff shortages that caused problems in the Servicing Department and lastly 'processes'. The heart of the system is the processes rather than the procedures.

'*Procedures*,' Alan read in ISO 8402, '*are a specified way to perform an activity whereas a process is defined as a set of inter-related resources and activities which transform inputs into outputs.*' Alan began to see the difference. Procedures define how an activity is to be performed and processes consist of all the other things that are needed, such as personnel, equipment and material, to deliver the desired output. Alan then thought of those quality manuals and, although it was some time ago, he was fairly sure that they didn't describe processes. All they covered were requirements and procedures.

Alan took up ISO 9001 once more and scanned the requirements. There was no mention of finance, human resources other than training and many other aspects for which his company had groups of people on the organization chart: Marketing, Tooling Design, Security etc. 'Why are these important functions missing?' thought Alan. 'Perhaps the standard is not applicable to these functions,' he mused. He then went back to Section 1 of the standard and read the paragraph on applicability. Here he found that the standard applies when a supplier's capability to design products needs to be demonstrated. 'Oh! That's interesting! It only applies when you have to demonstrate something. So if we don't need to demonstrate something to our customer then the standard won't apply to that. That's very interesting. It seems that the customer only requires things to be demonstrated if it directly affects what he buys and sure enough Security, Tool Design, Marketing is not something that we sell to our customers. So there really seems to be a difference between a quality assurance system as covered by ISO 9001 and a quality management system as covered by ISO 9004. Quality assurance seems to be about demonstrating that requirements are met and quality management seems to be the mechanism for achieving quality. But ISO 9001 is not specific as to the type of system we are required to establish. It simply uses the term *quality system*.' He then looked again at the ISO 8402 definition of a quality system and here it clearly stated that its purpose was quality management not quality assurance. 'I don't know who wrote these documents,' he thought, 'but they sure are confusing.'

Alan's story reveals several key messages fundamental to the design of effective quality systems:

- The importance of mechanisms that prevent failure and cause success
- The importance of education and training
- The purpose of documented practices
- That systems are the interconnection of related processes
- The need to explore the whole subject, not just skim the surface

These and many more aspects will be addressed in Chapter 2.

## The need for management systems

All organizations have some form of management system. In the smaller ones, it tends to take the form of a few written policies and instructions largely aimed at constraining the actions and decisions of the employees. These may address financial, operational, administration and personnel issues with forms used for requesting services, recording and reporting various activities and results. Many of the core operations remain undocumented. Accepted practice is passed down from the owner verbally to the employees and embedded into the folklore. Except for financial matters, work is carried out by passing verbal requests or memoranda to people who have learnt the way things are done in the organization. There is a lot of custom and practice and unwritten policy and people learn by example and from their mistakes.

This unstructured form of management system will survive for many years and will continue to be used until the burden of communicating everything verbally gets unworkable. Verbal instruction works when the message being given is relatively simple and only needs to be conveyed once. The size of the audience is not a factor as verbal messages can be conveyed effectively to thousands of people, as evidenced from the biblical texts and present day radio and television. When the people to whom the message is addressed are not present, the message needs to be documented in some form for them to read later. The message also needs documenting when it is complex and needs to be passed down to others or referred to frequently.

Verbal instruction cannot be replaced by written instruction as the verbal message conveys more than just words. The way the message is delivered, the body language, the emphasis and the environment in which it is delivered all play a part in conveying the message. Attitudes, ideas, beliefs, culture and behaviour are developed not out of documented messages alone but are enhanced from the power of the spoken word and visual images. The subliminal impact of television commercials has for years been a proven method of imparting ideas and beliefs.

The need for documented communication is no more evident than in the classic example from the First World War. The General passed down the message 'Send reinforcements we are going to advance.' By the time it reached the troops, having been passed from division to division, trench to trench, soldier to soldier, it emerged as – 'Send three and four pence we are going to a dance!' Thus the importance of written communication is illustrated if the original message is to reach its destination uncorrupted.

In organizations with unstructured management systems, memoranda and directives form the basis of written communication. When organizations become too large for the owner to convey all the instructions verbally, managers are employed to act on his/her behalf; this is where communication can become ambiguous unless means are used to ensure a uniform interpretation. Some owners will issue written directives, others will use team-building techniques and continue to convey messages verbally, confident that their disciples have understood and accepted the owner's vision and values and are following the creed. In the larger organizations, the managers document frequently-used practices so as to be able to rely on certain tasks being performed in the same way each time.

Every organization is different. Each will use a combination of verbal and written communications for work to be carried out and to obtain information on whether objectives are being met. Even with fully documented management systems, verbal communication remains a vital part of the system. However, unstructured communication, whether verbal or written, does not constitute a management system and, even in the well documented management system, verbal communication remains vital. It is interesting to note that with the advent of electronic mail, organizations have been flooded with all types of written messages, many of which are an unwanted distraction which do not, in fact, improve communication.

Whilst the primary purpose of a management system is to aid communication, it is not for that reason that organizations are installing formal management systems – at least they *seem* to have other motives.

The unstructured systems could not guarantee that customers would be supplied with products and services that consistently met their expectations. Unstructured systems are susceptible to unpredictable variation since the processes that cause the results can be driven by personal motives which may be at variance with the goals of the organization. Internal competition can cause friction and knowledge can be used to exert power over others, regardless of the overall effect on the organization's performance. The prevention of error is achieved by chance rather than by design. Success depends upon the influence of certain powerful people. Whilst the balance of power remains with these key people, the organization will deliver performance. When such organizations are successful they are unable to define why they are successful except by reason of the excellent people they employ. However, change any of these factors and the fragile system will break down. When such organizations fail they blame the people or the system without any evidence that either was the cause. When key people are absent or when new people join the organization, its performance is often adversely affected. When new products, services or processes are introduced, there is often a prolonged period during which any problems can be detected and resolved. Sometimes such problems remain unsolved, waiting for investment that somehow is never available, and success continues only because people compensate for these inadequacies by working long hours and doing things themselves rather than depending on others whose job it really is.

An attitude of 'If it's not broke don't fix it' prevails. An attitude of throwing resources at problems and firefighting also pervades the organization. Problems are fixed by focusing on symptoms rather than root causes. Without a structured system the fixes are never permanent and the problems eventually recur. Due to the informal compensating mechanisms the system appears effective, but remove those who compensate for the inadequacies and the system will break down. Communication breakdown is frequent and long working hours is a common characteristics. Even so, such organizations can survive for years and make a handsome profit. With little competition, there is no incentive to change the status quo. When competition emerges, cost cutting is often thought to be the solution. Cutting out indirect activities such as training, measurement and staff development without a thought to the consequences puts more strain on already overworked managers and staff. The organization becomes ever more dependent upon its scarce resources and when these too fall below the critical level, unrecoverable failure is inevitable.

Whilst this may seem an exaggerated description of organizations without structured management systems, it is not uncommon. Even organizations with formal management systems will have these problems if the system they have installed is not sufficiently robust. It should be recognized that a system is not merely a collection of loosely-arranged components but an integration of interrelated dynamic processes that rely on a number of factors, of which communication is but one. If this is ignored success will not be sustainable.

What is needed is a mechanism that does not rely on chance but a mechanism that will cause the rights things to be done and prevent the wrong things being done.

Where the production of products or information requires groups of people to operate processes that deliver consistent results, some form of management system is essential to cause this to happen. Even where the production process is intended to produce different results each time there is still a need for formal routines.

A painter often wants to produce a picture of original quality. The characteristics to be created in the painting are known only to the artist. However, the artist will prepare the canvas in the same way each time, use only certain materials mixed in a specific consistency and applied using a well-tried technique. The artist uses procedures and standards to attain the desired result but they are largely undocumented, except in text books. Success depends on the skill of the artist but is also dependent upon the quality of the materials used.

Design is a creative process in which the designer applies scientific theory to practical problems in order to produce workable solutions. One might think that a management system is unnecessary and may inhibit creativity but design is a process just like painting. However, unlike the painter, the designer works with knowledge rather than materials. The quality of the knowledge is vital: use the wrong data and the design won't work. The quality of materials, by way of contrast, is less important. Whether the design is represented on a match box, napkin or a piece of drafting paper is irrelevant for it to work.

Where there are no boundary conditions such as delivery targets or quality standards then there is little to manage. Work without requirements and targets is leisure, which is why we all need some leisure time, unconstrained by other people. For such activities we need no management system. For everything else we need to manage either time or quality and often both. If the artist is a member of a school of artists that is commissioned to produce work to defined schedules then some formal management system is needed to ensure that the prescribed delivery and quality targets are met. If the designer is commissioned to design a product to meet a customer requirement, then again a system is needed to ensure it happens. In business everything is constrained by requirements, targets, objectives and standards, otherwise businesses would not be businesses for long!

We need management systems when we have to organize and direct resources to achieve objectives, although some organizations can operate very successfully without formal management systems. The need for formal management systems arises where the production is repetitive either in the process that is executed or the results that are obtained and where such results are required to meet defined objectives. The numbers of people involved, the quantity to be produced, the type of product or service is irrelevant.

The popularity of ISO 9000 and the pursuit of ISO 9000 certification world-wide has created a demand for quality systems in order to give customers confidence that their quality requirements will be met. ISO 9000 certification has become a necessity for trade in some quarters and regardless of the advantages and disadvantages, organizations pursue ISO 9000 certification either because of competition or customer requirement. Rarely do organizations pursue ISO 9000 certification without some external pressure to do so. In such cases the need is survival in a competitive market rather than a means of managing the business. However, organizations are beginning to perceive that there are tangible benefits from a quality system beyond mere certification.

## The emergence of management systems

The first systems to be developed were likely to be financial management systems. Since economic performance is vital to all organizations, even non-profit organizations, the effective management of income and expenditure comes first. Whether the organization is very old or very new, a sole trader or multinational, these issues have to be formally controlled before bringing other activities under control. Coupled with the financial controls are the inventory controls for those organizations that use materials to provide revenue. Enterprises employing people need to bring their personnel management function under formal control so that the organization is furnished with adequate human resources. Since few organizations are not self-sustaining, the acquisition of materials, equipment, services etc. is a necessity, so these functions must be brought under control. Even so, the degree of control is usually limited to the transactions between departments: forecasts, budgets, purchase orders, material requisitions, recruitment requisitions etc. The operations of the departments that process these forms are documented in the form of simple instructions, supplemented by standing orders which are produced in response to a problem which has arisen and which the organization does not want to recur.

Anywhere in the organization where there are routine operations, one can find instructions designed to make practice consistent and limit the freedom of individuals. Starting with simple instructions, these documented routines grow into departmental procedures. Departmental procedures are most common in production departments where masses of information are processed through common processes or where materials are converted into saleable products through common processes. Insurance and car industries are but two examples.

Many successful organizations operate on a departmental basis with discrete operating procedures for each department. Where there are overlapping activities or interdepartmental transactions, common procedures are often developed.

## Standards for management systems

In recent years several standards have been developed in the field of management systems. One might say that the standard that got the ball rolling was BS 5750, introduced in 1979 and from which the ISO 9000 series was later developed. The UK continues to champion the development of system standards and following BS 5750 came BS 7750 on Environmental Management Systems. This eventually evolved into the ISO 14000 series. From the BSI stable also comes BS 7799 on Security Management, BS 8800 on Occupational Health and Safety Management Systems and BS 7000 on Design Management Systems. This is not a complete list, as I am sure other nations have developed similar standards. The single advantage of national and international standards is the fact that they are developed in conjunction with the users who bring together knowledge from many different sectors of our society. Unlike books published by one author (as this one is), standards represent the views of a group of people. With international standards, a group of people from different countries, with language barriers to overcome, agree on common standards that they would all be willing to adopt in their own countries. The net result is a document that has the agreement of the majority of participants (which is by no means everyone on standards committees world-wide) and hence will reflect the highest level at which agreement has been reached. This is certainly not the latest thinking on the subject, and is likely to be the lowest standard for many organizations. To obtain the latest knowledge one would need to seek out technical papers from practitioners and academics. Even in the time it takes to get a book published, the world has moved on since it was started and so a book, a paper, a standard is only a snapshot at a point in time.

Increasing pressure is being put on standards organizations to reduce the number of standards on management systems, primarily because of the attendant costs. The ISO 9000 series and ISO 14000 series are used for certification purposes and accreditation bodies presently insist that auditors of quality systems are not competent to conduct audits of environmental management systems and vice versa without special training and knowledge. If to this we add security, finance, personnel management, design management, project management and many more, firms foresee their costs going sky high, paying for an ever increasing number of certifications to keep ahead of the competition. There is, therefore, strong pressure for a single standard on management systems, integrating all of the best principles from the existing standards and providing a basis for single certification by a single certification body. Some companies have approached the problem head on by insisting that they have different certification bodies for different types of certification. It increases the costs, but keeps the certification bodies on their toes since there will be a tendency for them to become complacent were they to cover all certifications in a company. This situation is even more suspect when the certification body also carries out personnel and product certification in addition to management system certification. The people, the product and the process would all be certified by one body. Hopefully, different auditors would be employed, but if it were the same audit team doing all three, there would certainly be less objectivity. Such bodies will be keen to hang on to the contract and unethical practices may evolve to safeguard the business. What is most important: the certification body, money or ethics?

However, this book describes the systems engineering approach to quality system development using one of these standards (ISO 9001) but you could as easily use other standards. The technique is just as valid. If one were to build a management system using the methodology described in this book one would be able to extend the system to encompass the requirements of any other management system standard.

## Quality systems and the small business

A small firm is classified by the UK Department of Trade and Industry (DTI) as one employing between 1 and 50 people. The medium enterprise is one employing between 50 and 500 people. Small businesses account for over 95% of all businesses operating in the UK. The criterion for determining whether a quality system gives advantages is not, however, based on numbers of employees but on the nature of the business and how the products and services are supplied.

We should not forget that ISO 9000 is a means of obtaining confidence in the timely delivery of products and services which meet customer requirements. There are other ways by which clients can obtain such confidence; in fact, before ISO 9000 came along, clients obtained the necessary confidence from:

- Personal knowledge of the supplier
- The supplier's reputation in the community
- The results of surveys and audits of products and processes

The usual forms of business are the sole trader, the partnership, the limited company and the co-operative.

## The sole trader

The sole trader is the simplest form of business establishment. Where the sole trader operates within the local community, there is often personal contact between the trader and prospective customers. Further afield, personal face-to-face contact remains the preferred means of clinching a deal. In such cases the sole trader relies on reputation to gain work. Over 95% of business comes from referrals either from previous clients or from colleagues in a network. Whilst the trader operates on this basis a good reputation for quality work, gained by personal confidence, earns him/her a far better achievement than an ISO 9000 certificate. If and when the sole trader operates outside the boundaries where his/her reputation is known, potential clients are unable to distinguish between the reputable and unreputable suppliers and may use ISO 9000 as a means of selecting potential suppliers. A sole trader that has ISO 9000 certification may stand an equal chance of gaining business as those without the certificate, as it is potentially as valuable as a reference from several satisfied customers.

If you are a sole trader and have been asked to register your quality system then ask your customers what benefits they expect to obtain were you to do so. It could be that they misunderstand the principles of ISO 9000.

The need for the sole trader to have a quality system depends on:

- The degree of independence from external sources
- The degree to which the product or service depends on the skills and training of the individual trader
- The complexity of the work

Where the trader does not subcontract services, depend on others or purchase products that are subsequently passed to his/her client, then the client buys a product or service that is the result of the trader applying his/her own skills and knowledge. The client chooses to buy from an individual, not a company, and therefore the degree of confidence in the supply is based upon personal credentials rather than company capability. The quality of the service or product is a function of one person's efforts and his/her operating efficiency and effectiveness.

Efficiency is a function of the trader's self discipline and having a documented quality system would not improve self discipline in a one-person business as there would be no one else to cause the trader to comply with his/her policies and practices. Quality is achieved through a chain of processes. If all the processes are carried out by one person and this person lacks self discipline, there is no chain with strong and weak links – there is only one link. Without self discipline this link is weak and likely to break.

Effectiveness is a function of learning the right things to do acquired through training and experience. If the trader lacks the right skills, knowledge and self discipline, a quality system will not provide it. However, with self discipline, the trader would be more inclined to recognize weaknesses and improve performance, and a quality system may provide an approach that causes the trader to perform the right tasks in the right sequence.

A formal quality system would therefore provide little advantage to sole traders performing cleaning, typing or translation services. Formal documented quality systems are of little value to sole traders who provide nothing but their own skills and knowledge, such as independent consultants (although some are believed to have obtained ISO 9000 certification as a demonstration of commitment).

Where the sole trader uses external sources, depends upon others or involves a complex array of processes, a simple documented quality system may provide added confidence to clients by demonstrating that:

- Client requirements are documented and agreed before work commences.
- The products and services designs are verified and agreed with the client before commencing production or installation.
- The selection and acceptance of purchases are subject to formal controls.
- Equipment used is capable of the required accuracy and precision.

## Partnerships

Any business of more than one person requires:

- Effective communication to ensure that information is passed on intact to employees or partners.
- Common standards to ensure consistency regardless of who performs the work.

The nature of the work also has a significant effect on the need for a quality system.

Where the quality of the product/service can be verified by a single check prior to delivery, then in organizations of two or more people, no formal quality system is necessary, although ISO 9003 does provide a minimum level of assurance for customers of such organizations.

Where the quality of the product/service cannot be verified by end-product checks, a formal quality system would bring added benefits.

## The limited company and co-operative

These types of organizations may be similar in size to the sole trader or partnership. Some limited companies are created simply to gain tax benefits and financial independence and a co-operative could be as few as three people. In both cases the principles outlined for the sole trader and partnership apply.

The criterion is not size but relationships. Even in a company of 50 employees, the need for a quality system is dependent on relationships rather than numbers. Such a company may be run by the owner with drivers as employees (as in a taxi firm). All the employees do the same job and if the owner uses a computer-based accounting system and runs the communication system, there may be no need for a formal quality system. Customer satisfaction is established through the feedback mechanisms. A simple rule book could suffice to convey the company's code of conduct, legal issues, vehicle maintenance and fare-collection procedures. A regular review of performance and customer surveys would give adequate feedback to determine customer satisfaction. Even so, many taxi firms don't do this. I have used many taxis and never once been asked to comment on the service provided. A simple prepaid questionnaire handed to the customer by the taxi driver would suffice.

The larger the number of relationships, the greater the need for formal controls. Number of employees is therefore not the critical characteristic for determining whether a quality system is needed.

## Summary

In this chapter we followed the story of Alan, a sales person with a manufacturing company whose pursuit of quality was not helped by a management that did not understand the mechanics of quality management. Alan also explored ISO 9000 and came upon several inconsistencies, both in the series of standards and in its implementation by another company. The story illustrates the kind of obstacles that prevent companies from being 'world class'. They may be making a handsome profit, but they do not realize how much better they could become if they were to embrace a systems engineering approach to quality system development. Some insights into the origins of management systems and the reason for their existence was provided, together with a brief explanation of the standards around today that focus on management systems and create a case for developing a quality system based upon process management or systems engineering concepts. Finally, an overview of quality systems and the small business was provided, showing that quality systems are not practical for sole traders and that the need for formal quality systems increases as the number of relationships grows.

## Chapter 2

# Quality system principles

### The established definitions

A quality system is defined in ISO 8402 as:

> *Organizational structure, procedures, processes and resources needed to implement quality management.*

Quality management is defined as:

> *All activities of the overall management function that determine the quality policy, objectives and responsibilities, and implement them by means such as quality planning, quality control, quality assurance and quality improvement within the quality system.*

By contrast, ISO 14001 defines an environmental management system as:

> *That part of the overall management system which includes organizational structure, planning activities, responsibilities, practices and resources for developing, implementing, achieving, reviewing and maintaining the environmental policy.*

None of these definitions tells us much about what a system is, although they identify the components of a system (the organization, resources, processes etc.) but not what constitutes a system. There is a hint, as we are told it is something for implementing an aspect of management. The definition for an *environmental management system* comes closer as it tells us it is something for implementing a specific policy. In the 'quality' definitions there is some inconsistency as quality management includes the determination of the quality policy; therefore the quality system would embrace this as well as the implementation of the quality policy. In the 'environmental' definitions, the *system* excludes the determination of the policy.

Logic would seem to dictate that we create systems to achieve some objective, or deliver some result. Having stated our quality policy and our objectives, we need a 'system' to ensure the policy is implemented and the objectives achieved. Hence the relationship between the quality policy and the quality system is as illustrated in Figure 2.1.

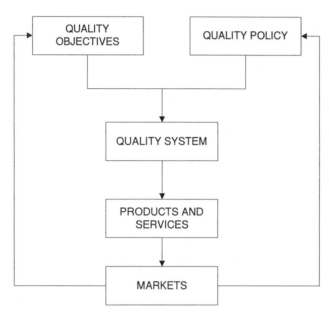

Figure 2.1 *Quality system cycle*

The model shows that the quality policy and objectives drive the quality system, which itself produces the products and services that are supplied to the markets. The market provides feedback to help improve the policy and objectives. If this cycle functions dynamically, it should enable the organization to become 'world class' in the quality of its products and services. The model therefore depicts a continual improvement cycle, a concept omitted from the ISO 8402 definition of a quality system but addressed by the EMS definition.

However, we need to appreciate a number of different aspects to lead us to a better understanding of the principles that underpin quality systems.

## Breakthrough and control

All managerial activity is directed at one of the following:

a) Breaking through into new levels of performance (that is, Breakthrough, alias creation of change).

b) Holding the resulting gains (that is, Control, alias prevention of change).

Juran[1] claims that there is a universal sequence through which managers achieve *breakthrough* and another universal sequence through which managers achieve *control*. These principles, which have been tested many times, can be illustrated through two diagrams, one showing the sequence of breakthrough (Figure 2.2) and the other showing the sequence of control (Figure 2.3).

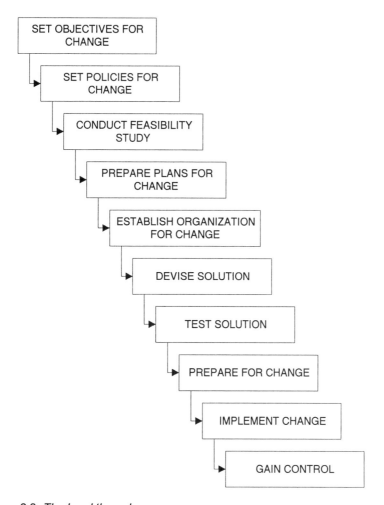

Figure 2.2 *The breakthrough sequence*

---

[1] From the preface of the first edition of *Managerial Breakthrough*, J. M. Juran (1964).

**24** Quality system principles

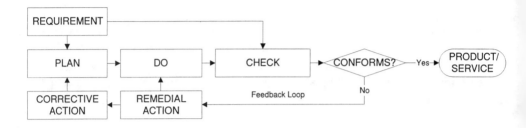

Figure 2.3 *The control sequence*

These principles are central to the management of quality and hence should be embodied in any quality system. The quality system should enable the organization to achieve breakthroughs under controlled conditions (that is, quality improvement) and to prevent change in the quality of business outputs (that is, quality control).

The relationship between breakthrough and control is illustrated in Figure 2.4.

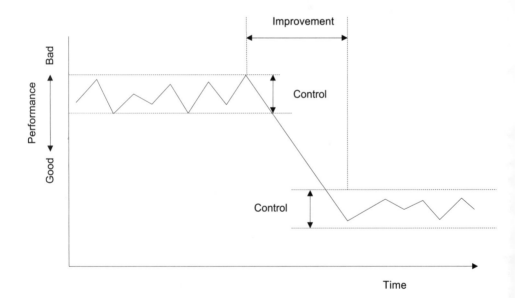

Figure 2.4 *Improvement and control relationships*

## The need for objective evidence

Customers often have no contact with the organization supplying the goods except through correspondence or telephone and have no way of checking for themselves that they will in fact receive what they require. Customers therefore need confidence that their requirements will be met by the organization with which they have placed their orders. Managers are likely to be remote from the activities that are carried out and although they are able to check that their policies and instructions are being followed, with so many things to check, they cannot spend all their time on this task as they have other responsibilities. Managers, therefore, need confidence that their policies and instructions are being implemented. Although specific solutions may differ, the methods employed to gain confidence in organizations or people have common aims. They all aim to acquire:

1 Knowledge of what is to be supplied. (This may be gained from the sales literature, contract or agreement.)

2 Knowledge of how the product or service is intended to be supplied. (This may be gained from the supplier's proposal or offer.)

3 Knowledge that the declared intentions will satisfy your requirements if met. (This may be gained from personal assessment or reliance on independent certifications.)

4 Knowledge that the declared intentions are actually being followed. (This may be gained by personal assessment or reliance on independent audits.)

5 Knowledge that the products and services meet your requirements. (This may be gained by personal assessment or reliance on independent audits.)

The manager may possess the knowledge in points 1, 2 and 3, as this information may have been supplied to the people charged with carrying out the work. The customer, on the other hand, may only possess the knowledge in point 1 and sometimes even this may be incomplete. Knowledge is acquired through objective evidence, either by witnessing events or by having documentary evidence of intentions and achievements. It is the validity of the objective evidence that provides the confidence. Seeing evidence that is subsequently proven to be false weakens your confidence. Relying on hearsay evidence that is later proven to be false also weakens your confidence. Having the independently validated objective evidence in front of you strengthens your confidence. Hence the need for documentation that demonstrates intentions and for records that demonstrate achievements.

## Quality is a result of the business

There are several results of a business that need to be managed, otherwise a failure in any one of them could jeopardize the business. They are quality, environment, health, safety, human relations and profit. Product quality results from knowing your customer needs and expectations and faithfully meeting them consistently within an organization that promotes leadership and continual improvement as its core values. Careless design of products and processes can result in a harmful effect on the environment; it can also result in harm to personnel, either by causing injury or by damaging the health of those making the product as well as of those using it. Negative cultural traits in an organization will result in poor human relations with lost time, inefficiencies, strikes, etc. Poor decision making and poor control result in low profit. Product quality results from the behaviour of people in an organization. Even when the policies and procedures will enable people to do the right things, the core values may work against this and cause conflict.

It is therefore not enough to focus on one of these results. Anyone could be successful in one result area at the expense of the others. Managers can make a profit but what value is it when the consequences are lower standards of product quality, safety, health and environment? Managers need to pay attention to all of these factors.

## Customer needs and expectations

In managing anything we need some objectives. Managing without objectives is coping rather than managing. The need for management arises because we wish to achieve something. It could merely be to utilize time effectively, in which case we would need to know how much time to spend on each task we were asked to do and judge whether such time would be wisely spent. We would then decide whether or not to spend the time. When the task is done we account for the time spent and, over a prolonged period, adjust our use of time accordingly so that we feel it is being utilized effectively. The approach is no different to managing anything else except that in addition to time we may have other factors to manage such as people, materials, space and money. In managing quality we need to know what is meant by quality and to get an answer to that question we look to our customers. We need to establish:

- Who and where are our customers?
- What are our customer needs and expectations?
- What customer needs and expectations are currently unsatisfied?

In many quality systems these simple questions have not been answered or, if they have, the answers have not been agreed and documented among the management. It is from the customer needs and expectations that quality policy and quality objectives should be derived.

## Failure prevention system

The basic goal of quality management is the elimination of failure; eliminating failure both in the concept and in the reality of our products, services and processes. In an ideal world, if we could design products, services and processes that could not fail we would have achieved the ultimate goal. Failure means not only that products, services and processes would not fail to fulfil their function but that their function is also exactly what customers require. A gold-plated mousetrap that does not fail is not a success if no one needs a gold-plated mousetrap!

We have only to look at the introductory clauses of ISO 9001 to find that the aim of the requirements is to achieve customer satisfaction by prevention of nonconformities. Hence quality management is a means for planning, organizing and controlling the prevention of failure. All the tools and techniques that are used in quality management serve to improve our ability to succeed in our pursuit of excellence. Excellence can be characterized by the absence of failure.

Quality does not appear by chance or if it does it may not be repeated. One has to design quality into the products and services. It has often been said that one cannot 'inspect' quality into a product. A product remains the same after inspection as it did before, so no amount of inspection will change the quality of the product. What inspection does, however, is to measure quality in a way that allows us to make decisions on whether to release a piece of work. Work that passes inspection should be quality work but, unfortunately, inspection is not 100% reliable. Most inspection relies on human judgement and human judgement can be affected by many factors – some of which are outside our control, such as the private life of the inspector! We may fail to predict the effect that our decisions have on others. Sometimes we go to great lengths in preparing organization changes and find to our surprise that we neglect something or underestimate the effect of something. Work that passes inspection is only quality work if the acceptance criteria used by the inspector reflect customer needs and expectations. We therefore need means other than inspection to deliver quality products. It is costly anyway to rely only on inspection to detect failures. We have to adopt practices that enable us to prevent failures from occurring.

When we know who our customers are, what they expect and how we operate, we can then plan and organize to achieve quality. We need to look at our processes and ask ourselves what could prevent us causing satisfaction or what could dissatisfy customers.

Most planning does include provision for failure detection but relies heavily on trained personnel knowing the right things to do. We need to analyse our processes and determine what could go wrong, anticipate the obstacles one might encounter by asking 'What if?' at every stage in the process, then put in place provisions to avoid failure.

However, imagine all your efforts being put into a task, detecting and correcting all the errors and then finding out that the product of your labours was not wanted by your customers. It is not enough to do things right, one has to do the right things as well. Your plans need to define the right things to be done without relying on chance.

The management of quality involves many aspects of an organization. In essence quality management is concerned with the failure potential of processes, products and services, as stated previously. Organizations comprise many functions and all must be essential for the organization to function efficiently and effectively. It follows therefore that if any function fails to perform, there will be a corresponding detrimental effect on the organization. Whether this failure has any effect on the products and services offered for sale depends on the time taken for the effect to be damaging. Some failures have an immediate effect where they contribute directly to the supply of products and services. Others have a long-term effect where their contribution is indirect, such as the behavioural aspects. A failure in a support function, such as office cleaning, may not affect anything initially but if the office remains unclean for a prolonged period it will begin to have a detrimental affect by modifying the behaviour of those who work in such an environment.

## Causing success

Merely preventing nonconformity is not enough. A system that appears to prevent nonconformity may not cause conformity. Some activities may occur by chance or be informal and may be the ones actually preventing nonconformity. If an organization achieves its quality objectives but not through the established quality system then that system is not effective. When seeking to eliminate failure we search for the root cause of problems and attempt to eliminate them. This search will discover events that arose due to weaknesses in the system. By eliminating these weaknesses we strengthen the system.

Sometimes we operate a process a hundred times and never experience a failure. Then one day our expectations are not met: the product is not right, the job is late, the product fails – not when it is under our control but when 50,000 of them are in service. 'What went wrong?' you ask. 'We have never had a failure like this before.' All the formal controls designed to prevent failure failed – or did they? Success may be a product of both formal and informal methods. People may be doing things that only they know how to do so that when they no longer apply their informal methods, the process fails. The problem with informal methods is that one never knows how important they are until they cease to be applied. To rely on a task achieving the same result no matter who does it requires a strategy that causes the right things to be done. It would seem therefore that quality systems should cause an organization to achieve its quality objectives as well as prevent failure.

## Using foresight

Many companies install quality systems and forget that it is quality they intend to manage. The third-party assessment process seems to concentrate on conformity and not on quality. It is assumed that if people follow procedures they will delivery quality products and services. This is not so. Unless the procedures are designed to work together to deliver quality so that product or information passes through controlled processes, quality, if achieved at all, will be achieved by chance and not by design.

In the design of products and processes we can build in features that prevent failure and where this is not possible, due to other constraints, we can install failure alarms or fail-safe devices. If we can't avoid failure we need to remove its source or control its effect. To do this requires knowledge of failure mechanics. We should establish:

- The potential for failure
- What its effects would be
- How severe it would be
- What the cause of failure would be
- The likelihood of its occurrence
- The likelihood of its detection
- What we could do to eliminate the cause or control its effect

Such an analysis gives foresight to managers. However, we are blessed with hindsight. It is a wonderful gift. If the architects of the world's greatest disasters had possessed foresight many lives might have been saved!

## Error detection and correction

Even if we have put in place the failure prevention provisions, this is not enough. Many failures cannot be prevented as their characteristics cannot be controlled beforehand. Human error is one such characteristic and needs detection techniques to be discovered. We therefore need to install the sensors to detect whether we are achieving our goal and put in place provisions to measure customer satisfaction. The control sequence in Figure 2.3 illustrates the concept.

In managing quality we have to know what causes success and what causes failure. We cannot operate blindly. Have we got control over quality or are we observers who sit in offices and wait for the reports and then take action? True, it is impossible to have a handle on everything that happens in the organization. We do not need to control by looking over the shoulder of every operator. We can control quality by preventing failure before the event, by preventing failure during the event or by correcting failure after the event and before our quality standards are breached. In some cases the risks of failure are so high that we have to watch very closely and constantly check systems. Imagine if NASA were to rely entirely on the controls exercised before the vehicle was placed on the launch pad, taking everything as being done correctly and not running through the pre-launch checks. Many times a spaceshuttle launch has had to be abandoned because of a fault being detected. Even with more routine operations, such as airline operations, pilots carry out checks before take-off as some unforeseen event could have occurred. The basis for this is not that they don't have their operations under control but there is a high risk of external forces outside their direct control being present and changing a previously checked parameter. In any operation, we need to anticipate what could go wrong and take appropriate measures to detect the presence of risk before proceeding.

## Quality is achieved through a chain of processes

All work is a process. Work is not done by passing instructions from managers to staff, vertically upwards and downwards in the organization. Work is carried out horizontally with responsibility passing upwards and authority passing downwards, as illustrated in Figure 2.5.

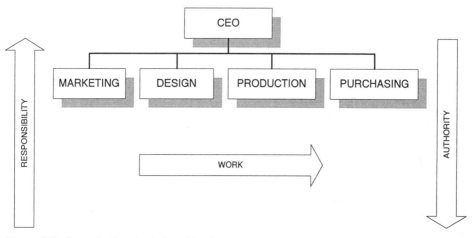

Figure 2.5 *Organizational relationships to work*

We need to know how we conduct our business from the identification of customer needs and expectations up to the delivery of the product or service and beyond, through after-sales services. If we don't know how we process orders, make purchases, plan production, install products etc. we will never understand what causes success or failure. It is not a matter of knowing what every function does in the organization but a matter of how these functions work together to deliver the product or service. We need knowledge of the planning activities, the producing activities, the controlling activities and the supporting activities and how they interact. More than anything else, we need to know the sequence of steps that are taken from input to output so that we have a complete picture. Without a complete picture there may be some activities being performed that are crucial to success and, if not accounted for, could result in failure should they be not done. Ever heard the cry 'I didn't know you did that!' or 'Had I known that was important I would have told you.' Such remarks often follow a disaster. These processes are linked together in the chain so that the failure of one link could affect the end result. More often than not, such failures are detected by adjacent functions that compensate for the errors. However, one must not rely on such informal controls. So we need to know all the activities in the chain that lead to, and subsequently cause, customer satisfaction.

## Behavioural system

Quality management is both a technical subject and a behavioural subject. A quality system includes people, their interrelationships and hence their behaviour. The behavioural system is therefore part of the quality system. The behaviour of people in an organization is developed by the culture within the organization. One can define some very practical policies and procedures but if the culture is not conducive to encouraging staff to use them, the desired output will not materialize. Quality results not only from using the right tools, the right processes etc. but also approaching the task with the right attitude. A person may be given everything needed to do an excellent job but, lacking motivation, he/she will invariably fail to meet the organization's expectations.

Culture is shaped by the core values expressed by the management. Without shared values no quality system will fulfil its purpose. Behavioural systems are often undocumented. To do so would expose the inequality, unethical behaviour and power system within the management structure. The simple yet profound requirement in ISO 9001 for the supplier to define and document its commitment to quality focuses on this difficult aspect of the quality system. A number of factors relating to the behavioural aspects are dealt with in Chapter 5, which covers such things as trust, ethics, integrity and respect. These and other behavioural attributes shape the organization's values that are part of the quality system.

The absence of core values to form a positive behaviour may not have an immediate effect, as individuals will operate to their own personal values. When these in turn conflict with the organization's values, the individual resents being forced to comply and eventually adopts the values of the majority or leaves to find a more suitable company to work for. Normally, such problems are remedied before they have an adverse affect and are not usually part of a management system. If a Total Quality Management philosophy is to be adopted then every function in the organization, regardless of the magnitude of its effect on processes, products and services, is brought into the system. ISO 9000 only addresses those functions that contribute directly to the sale of products and services to customers. The difference is that ISO 9000 and other standards used in a regulatory manner are not concerned with an organization's efficiency or effectiveness in delivering profit or nurturing the values that determine the behaviour of the people who make decisions. Organizations need to pursue standards beyond the published standards to become 'world class'!

## Technical or administrative system

There has been a perception in the service industries that ISO 9000 quality systems only deal with the procedural aspects of a service and not the professional aspects[2]. The rise in popularity of ISO 9000 has created some unhelpful messages, such as the 'document what you do' strategy. For instance, in a medical practice the ISO 9000 quality system is often used only for processing patients and not for the medical treatment[3]. In legal practices, the quality system again has been focused only on the administrative aspects and not the legal issues[4]. However, in management consultancy, the systems are intended to focus also on the quality of advice offered[5].

The argument for this is that there are professional bodies that deal with the professional side of the business. In other words, the quality system addresses the non-technical issues and the profession the technical issues. This is not quality management. The quality of the service depends upon both the technical and non-technical aspects of the service. A patient who is given the wrong advice would remain dissatisfied even if their papers were in order or if they were given prompt and courteous attention and informed of the decision promptly. To achieve quality one has to consider both the product and the service. A faulty product delivered on time within budget and with a smile remains a faulty product of no use to the customer. Likewise, a quality product delivered on time but with a grudge will not encourage the customer to return.

---

[2] *Quality World* (July 1997).
[3] ibid.
[4] *Quality News* (March 1991), with reference to the first law firm to achieve BS 5750 certification. It is stated that the certification is not for the quality of the advice tendered by the firm but acknowledges the implementation of the firm's quality and training policies and objectives and its management system.
[5] A guide to the interpretation of ISO 9001 for Management Consultany (1990) states that 'product' is any materials, advice or service rendered to the client.

## Key messages

- The basic goal of quality management is the elimination of failure.

- Quality results from knowing your customer needs and expectations and faithfully meeting them consistently within an organization that promotes leadership and continual improvement in its core values.

- Quality results from the behaviour of people in an organization. Even when the policies and procedures will enable people to do the right things, the core values may work against this and cause conflict.

- It is from the customer needs and expectations that the quality policy and quality objectives are derived.

- The problem with informal methods is that one never knows how important they are until they cease to be applied.

- It is not a matter of knowing what every function does in the organization, but a matter of how these functions work together to deliver the product or service.

- Without shared values, no quality system will fulfil its purpose.

- We invent systems to achieve a certain objective, deliver a certain result.

- A faulty product delivered on time, within budget and with a smile remains a faulty product of no use to the customer.

- Organizations need to pursue standards beyond the published standards to become world class.

## Quality systems or quality management systems

In ISO 9000, the terms *quality system* and *quality management system* are interchangeable. There is no difference. In the 1987 versions, both terms were used but, in the update in 1994, the term *quality system* was adopted throughout. A quality system could be perceived as a system to achieve quality, whereas a quality management system could be a system to achieve quality management. The term *quality management* does have some historic baggage that can tend to hinder communication. It can be perceived as the role of the quality manager rather than the role of all the managers. Hence the term 'quality management system' could send out the wrong signals and result in the system being perceived to apply only to the quality manager and his/her staff. Quality management is often referred to as a function of the business and hence gets assigned to a specific manager. By dropping the word 'management' the likelihood of misconception is reduced, but not eliminated. In organizations that historically have not had quality departments these problems are less likely to arise.

## Quality systems and Total Quality Management (TQM)

*Total Quality Management* (TQM) is not a technique, a discipline or a standard. It is a set of principles. The management of quality can be viewed in three different dimensions[6]: the business dimension, the organization dimension and the product dimension. ISO 9000 only focuses on the product dimension. By perceiving quality to extend in these dimensions rather than being limited to the product dimension, one is embracing TQM. It is *total* in that *every* facet of the business has an impact on the quality of its products and services.

ISO 8402 defines TQM as *a management approach of an organization centred on quality, based on the participation of all its members and aiming at long-term success through customer satisfaction, and benefits to all members of the organization and to society.*

This concept suggests that quality is not limited to being an attribute of a product or service but an attribute of everything an organization does in society. There is the quality of products and services and the quality of the decisions taken by the organization that affect itself, its customers and society. The quality of products and services comprises the quality of their design, the quality of conformance to the design, the quality of installation, and the quality of servicing. We should also remember that the product is not the only thing we buy but also the packaging and the literature supplied with it. Another aspect to take into account is the quality of the service associated with the product, such as the pre-sales service and the after-sales service.

All organizations, whether profit or non-profit making, exist to provide something for others. The motive of the profit-making organizations is to create customers for their products and service and in so doing yield a profit that can be shared with their shareholders or owners. The motive of the non-profit-making organizations is to benefit others without making a loss. Every person, piece of equipment, building etc. exists to serve the objectives of the organization. It follows therefore that every activity carried out must have a direct or indirect effect on the quality of the business outputs. If you can find an activity that does not have such an effect then you ought to question the reason for its

---

[6] See also *ISO 9000 Quality Systems Handbook*, page 11.

existence! ISO 9000 tends to focus on those activities that have a direct effect on the product or service, whereas TQM brings the whole organization into focus. This would therefore imply that ISO 9000 is but one aspect of TQM. Other management system standards focus on different aspects of TQM, such as the environment, health and safety, security or human resources. Although other approaches can be viewed as complementary to quality management, they are in fact an aspect of quality management. Quality management is often perceived as a means of managing operations so as to prevent failure and indeed many of the tools and techniques used in TQM do in fact serve this purpose; for example Pareto Analysis[7], FMEA and SPC[8]. However, quality management is not just a set of tools to prevent failure. It is also a means for causing success and hence must embody the tools for creating organizations, for creating markets, for creating products and services. Some of the tools used in TQM also serve this purpose, such as Benchmarking, Quality Function Deployment and Design of Experiments. However, TQM is not about tools, it is about culture. Some of the tools (such as, team building, critical success factors, goals and targets) are used to form, sustain or change the culture.

## Relative position of systems in TQM

So where does the quality system fit within TQM? It should sit at the heart of TQM. Whether or not an organization has a 'documented' system, it will still have some form of system: *'The way that things are done around here!'* Every organization is different, even those that provide the same kind of service (e.g. dentists, banks, filling stations) Although the processes will be almost identical in each, the systems will differ owing to the differing personalities of those involved. It is possible to import a documented system from elsewhere, such as in a dental practice where the administration, appointments system and rule and standards of hygiene might be the same. But this suggests that the word *system* consists merely of the documented methods and no more. ISO 9000 is not about a system of documentation, it is about documented systems for achieving quality. ISO 8402 defines a quality system as the *organization structure, processes, procedures and resources needed to implement quality management.* So let's examine each of these further.

### *Organization structure*

To many this means an organization chart or several charts that depict the various positions and their interrelationships. But this is where it stops. Is this all there is to an organization structure – a piece of paper with boxes containing names? No! The organization structure is far more than this. As David Packard remarks[9]: 'The way the organization is structured affects individual motivation and performance'.

---

[7] After Wilfredo Pareto. See *Quality Control Handbook*, J. M. Juran.
[8] Failure Mode Effects Analysis. Statistical Process Control.
[9] *The HP Way*, David Packard (Harper Collins, 1995).

Consider two organization charts, one for a private company that maintains road vehicles and another for a unit of the military that maintains road vehicles. They may well look alike on paper but in practice they might be quite different. The private company may operate in a way that provides employees the freedom to innovate, to participate in decisions, to discuss matters directly with the MD, to work flexible hours. The military unit may have none of these freedoms; every employee may have to obey the instructions handed down from above without question. Each is appropriate to its environment but neither is shown on the chart.

If the quality system is to comprise the organization structure then it must consist of more than a chart. There should be a description of the organizational policies that are employed to foster individual motivation and performance. These could cover:

- Division of work within the organization and its component parts

- Selection of personnel for appointments and the promotion policies

- Defining and communicating responsibility and authority

- Staff involvement in policy decisions

- The way policies and instructions are issued and feedback obtained

- The use that is made of staff suggestions

- Resourcing of projects and redeployment of surplus resources

- Employee benefits, share options, sick pay, insurance schemes etc.

- How managers obtain the confidence of the workforce

- How managers show their trust in the workforce

- Policies on working hours, holidays, absence etc.

The vision and values to which the organization subscribes and the ways in which these are embedded in the culture and periodically reinforced will also be important. What all this relates to is, of course, the style of management and, therefore, the organization structure becomes a reflection of management style. To reflect this accurately, the description of the organization structure should cover all significant facets of the management style, such as those above.

## Processes

Processes are the means to transform inputs into value added outputs. They include all the materials, machines, environment, personnel, documentation and techniques needed to convert given inputs into required outputs.

By taking a modified *Plan Do Check Act* (PDCA)[10] cycle and adding the element numbers of ISO 9001, one can allocate most of the elements, as illustrated in Figure 2.6. Here the output is the specific product or service of the organization. The contract or customer order is the input requirement, the PLAN function consists of the documented quality system plus the design and production/installation/servicing plans. The DO function is the activity of design, purchasing, production, installation and servicing for specific products and services. The CHECK function not only consists of the inspection elements of ISO 9001 but the process-monitoring aspects of clause 4.9 and the measuring equipment. Statistical techniques (clause 4.20) would also be applied in carrying out the process monitoring and inspections. The ACT function consists of two components: action on the nonconformity (Remedial Action, covered by clause 4.13) and action to prevent recurrence (Corrective Action, covered by clause 4.14).

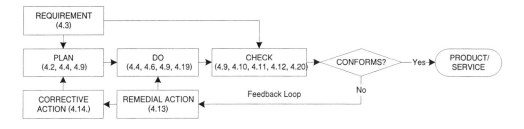

Figure 2.6 *Operational processes*

If we now consider the whole organization rather than specific products and services, then we can redraw the PDCA cycle and apply the other elements of ISO 9001, as in Figure 2.7.

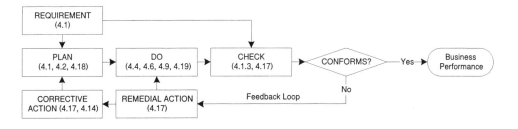

Figure 2.7 *Management processes*

---

[10] A concept articulated by Edwards Deming.

The input requirement is the quality policy and objectives. The PLAN function is the documented quality system, the resources and training needs. The DO function is the activity of design, purchasing, production, installation and servicing for all products and services. The CHECK function is the internal audit element with the ACT function again being split between the remedial action requirement in clause 4.17 and the corrective action requirement of 4.14.

## *Procedures*

A procedure is simply a way of proceeding to accomplish a task. It can be a few steps or a book full of instructions. How do we proceed to do anything? Step by step, rather than in huge jumps. Any document that lays out a way of doing something step by step is a procedure. We often view procedures as documents but by taking a new perspective and using a bit of lateral thinking. For a procedure to work effectively it must include the intangible as well as the tangible and hence include the way a task is approached by an individual – their behaviour as well as the effectiveness of any training and education. The smooth operation of procedures depends as much upon the people using them as it does upon the words in the document in which the procedures are described.

For example, in Figure 2.8 there is a sequence of steps. The responsibility has been assigned and the method to be used to carry out each step defined. In theory, the output should meet the requirements but in practice there is much more influencing work in this procedure than these steps. There are unstated interfaces with other procedures; equipment maintenance and document control for instance; interfaces with policies such as the quality policy, environmental policy; and probably the most important interface is the human relationship between the operators and the supervisor. Should this relationship be negative, there is likely to be a high reject rate in the output – if it is positive then a low reject rate is likely. How the supervisor passes the requirements to the operator between steps 1 and 2 has a significant bearing on how well the requirements will be understood. How the supervisor passes back the results of the inspections will also have a significant bearing on how co-operative the operator feels about fixing the errors – the requirements may well be cited as being the source of the errors. The operator carrying out step 5 is passed a piece of work to move to another operation. He/she does as he/she is told! Moving product all day long can be pretty tedious. Procedure OP015 referenced in Figure 2.8 may only describe the steps and precautions, not the manner in which product is moved.

## *Resources*

The system includes the resources to implement quality management. Resources include personnel, materials, equipment, facilities, plant, space, finance and time. To describe resources the system would have to describe how each resource requirement is identified and the identified resources acquired. A system without adequate resources will not function. If the managers do not allow sufficient time for the issued procedures to be carried out, the system will be ineffective. If there are insufficient or inadequate tools, equipment,

Quality system principles  **39**

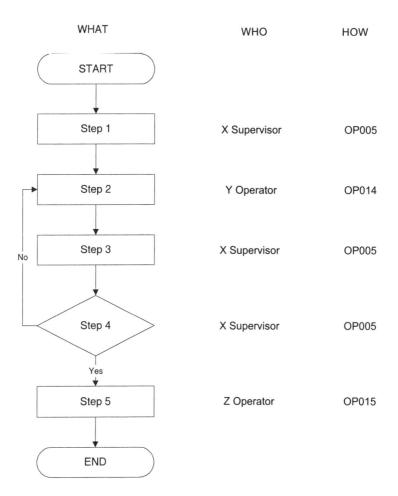

Figure 2.8 *Simple procedure*

staff and space for the activities described by the procedures then the system is ineffective. If the staff are not equipped with the education, training and motivation needed then the system is not effective.

Finally, returning to the people factor, we could view motivation as an element of resources and consider that the organization structure may well affect individual motivation. Everyone has the ability to be motivated. The secret is to create the right environment to extract that drive.

If the system is purely the documentation or the tools to achieve quality, then it does not sit at the heart of TQM; it sits at one side, as illustrated in John Oakland's model[11] (Figure 2.9).

---
[11] *Total Quality Management*, John Oakland (Butterworth-Heinemann, 1993), page 435.

**40**   Quality system principles

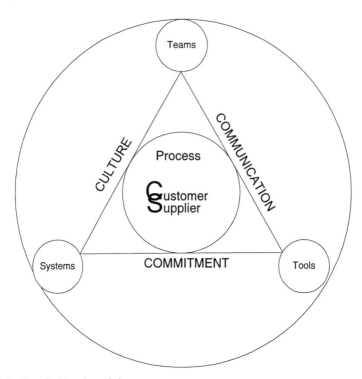

Figure 2.9 *The Oakland model*

But if the system comprises resources, processes, organization structure and procedures (as required by ISO 8402), it will include the teams and the tools in Oakland's model. Outside the system are customers, suppliers, science and technology. Science and technology is outside the system simply because it is a variable that exists alone, knowledge that is used in conjunction with the system to produce the desired products and services.

People may not like being regarded as part of the system and included as a resource. People are free to join the system, be a part of it or leave. When one joins a club, one is governed by the club rules so it is not the human being that is part of the system but the individual's behaviour and their actions and decisions. Once in the system, it limits the behaviour of those within it for the good of the organization.

An alternative model of TQM is illustrated in Figure 2.10.

*Quality system principles* **41**

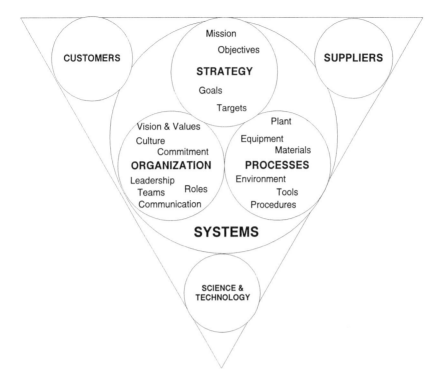

Figure 2.10 *System relationships*

Science and technology are at the bottom of the model because they support everything we do. Without the science and technology to do things, no amount of strategy, organization and processes would achieve very much. However, don't think of science as being academic and technology as being computers, machinery and the like. Science and technology are merely a means to understand and accomplish a task – their application can lead to just as many failures as successes. The key to the successful use of science and technology is through an effective system. In the model (Figure 2.10) only three elements are shown as comprising the system, not the four in the ISO 8402 definition. This is because processes can be considered as including resources, for without them they are not processes, merely descriptions of processes. Procedures are also a part of processes, for without them the outputs would be the result of trial and error methods. Systems are at the centre and interface with the customer's and supplier's systems.

## TQM relationships to ISO 9000

Let us now look at other relationships between the system and TQM. As shown in Figure 2.11, each of the factors included in the TQM approach is dealt with in ISO 9001, although not as detailed as in other texts. A detailed analysis of these relationships may be found in the *ISO 9000 Quality Systems Handbook* (Third Edition).

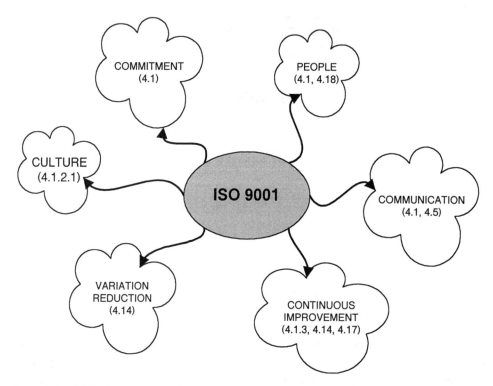

Figure 2.11 *TQM elements in ISO 9001*

## Summary

In this chapter we have examined the definitions of a quality system, the concepts, the basic terms and how they relate to each other. Instead of defining a quality system as a composite of many components, a more appropriate definition might be the integration of interconnected business processes which collectively cause the fulfilment of customer satisfaction and prevent the supply of nonconforming products and services.

We have also identified relationships between quality systems and TQM and shown there to be many common elements. However, if one takes a very limited view of ISO 9000 and ignores ISO 8402, one might have some difficulty in relating ISO 9000 to TQM. Only by taking a new perspective will one see the connections and realise that a quality system is not just a set of documents but the dynamic entity which includes people, values, continuous improvement and all those other elements of TQM. All you have to do is open your eyes and see more deeply into the words. Remember that the fewer the words in a standard, the more important each one becomes.

A TQM programme needs to have at its heart a dynamic quality system that is used to drive continual improvement. Even if you limit your system to the 20 elements of ISO 9001, it should still drive continual improvement. Let us take a simple example.

After running the quality system for some six months we are able to collect and analyse quality data on nonconformities detected at several stages in the life cycle of a particular range of products. Applying our corrective action procedures (clause 4.14.2 of ISO 9001), we identify the causes of the nonconformities and to eliminate the cause requires changes in product specifications. We then ask ourselves why these changes were necessary and find that the specifications were released before completion of the design verification tests. We investigate further and find that the verification test was delayed due to the unavailability of test equipment, which was being used by another department. When we question the manager concerned it is revealed that he neglected to inform the test department when the equipment would be required. We conclude that the root cause on this occasion is poor development planning.

Six months later we analyse the data again and once more discover a similar problem with the same group of designers. The manager has failed yet again to plan development properly. So we convince him that some retraining in project planning may help him plan projects more effectively. A few months later we discover that the scheduled training had not been carried out as the Development Director had diverted the manager to another project where there was a resource shortage. Once again the manager failed to plan the development of this project properly; however, we now had a director who appeared uncommitted to quality. On further investigation we discover that this particular director had no training programmes for his managers and had not included training in the divisional budgets. The budgets had been approved by the CEO although the budgeting procedure clearly stated that training was to be included. We have revealed a serious problem with commitment and explore further to find that the CEO is not expressing the importance of the quality policy with his managers. If this continues the whole company may slowly deteriorate.

A knight in shining armour comes to the rescue. No! Not the Quality Manager but the Finance Manager who just happened to sit in on the quality awareness training and has been reading about quality in the financial journals she receives. After bending the ear of the CEO, a new quality drive emerges and everyone is included. The quality system has attracted many followers.

In conclusion, the management of quality starts by recognising that quality is a result that is affected by the core values of the organization. Quality is achieved through a chain of processes that focus on customer needs and which provide for failure prevention, detection and correction, thus resulting in customer satisfaction with the products and services supplied.

## Chapter 3

# Preparing for project launch

## Project overview

The development of a quality system should be treated as a project and not as a job for someone when time can be found – because the time needed will never be found. By treating it as a project, the work will progress through distinct phases[1]. A project has a start and an end. However, the end event is only a beginning for the users of the quality system and therefore this book does not stop at system evaluation. Each of these phases is outlined below.

### Phase 1 – Exploration
The necessity for a quality system arises when one person feels motivated to change the status quo (for example, the pressure of customers, the level of firefighting, complaints, etc. may need to be changed). Managers then need to be convinced that a change is necessary before proceeding further and before measures are taken to achieve the necessary change in attitudes.

### Phase 2 – Project feasibility
Before assembling the resources of the company to achieve a breakthrough, a feasibility study is undertaken in order to establish where and to what end the resources are to be applied.

### Phase 3 – Project planning
Having obtained support for the project, a plan of action is produced and resources organized so that the effort is directed along a course that will lead to a successful conclusion.

---

[1] See *Managerial Breakthrough*, J. M. Juran (1964 – revised 1995).

## Phase 4 – System design
There follows a period in which a view of the business is established – the business processes are defined and the documentation needs and areas requiring change are identified. This is the most crucial stage as it lays the foundation for the system.

## Phase 5 – Documentation development
Once the documentation needs are identified, the new and existing practices are documented. To overcome resistance to change the targeted users of the documents participate in their development and are involved in the review and approval process before implementation.

## Phase 6 – System implementation
Once the documentation has been approved and resistance to change overcome, a plan of implementation is produced, training undertaken and documented practices released for use.

## Phase 7 – System evaluation
When the new practices are being implemented the effectiveness of the system is evaluated on a continuous basis as a safeguard against a return to the old practices. It is also at this stage that external assessments are conducted and the system is certified against an international standard.

## Phase 8 – System maintenance proving
The Project Team will continue their vigilance following certification in order to keep practices in line with policy, keep records up to date and prove the effectiveness of the maintenance elements of the system. Changes will be made in response to problems and to keep pace with changes in the organization and in technology.

## Phase 9 – System improvement proving
Improvements in the policies and practices wil be pursued in order to prove the improvement elements of the system. As better ways of doing things emerge and customer needs and expectations change, a programme of continual improvement will be undertaken in order to drive the organization towards world class.

In this chapter, the first two phases will be examined: Exploration and Project Feasibility.

# Exploration

Exploration can be divided into two stages – identifying the need and proving the need.

## Identifying the need

The need for a formal quality system can arise through one person's experience, as it did for Alan in Chapter 1. Let us summarize these:

a) An increase in the number of customer complaints

b) Evidence that the existing system is not working in the best interests of the company

c) An observation that others were formalizing their quality systems

There are other reasons that one might come across:

a) A simple idea that performance can be improved

b) Direct enquiries from existing customers

c) An observed decline in demand for existing products and services

d) An observation that competitors are widening the trade gap

e) Condition of trading with certain countries (e.g. European Union)

f) Condition of trading with certain sectors of industry (e.g. Automotive, Pharmaceuticals, Aerospace)

g) Decline in company profits and economic performance

At this stage nothing has been quantified. How much, how long, who will be affected – all remain uncertain. There will be no change unless there is first an advocate of change but for the idea to reach maturity, attitudes will probably have to change. The need has to be proven to others.

## Proving the need

For Alan in Chapter 1, the need was proven, but his experience was personal and perhaps not shared by everyone. It is safe to assume that everyone will want fewer customer complaints, better quality, better delivery and lower costs but they may all differ in the approach they would be willing to take to achieve such improvements.

Installing a quality system is not something that one can do in the back office. It has to be done in the open with the participation of every manager and the CEO (not their consent or involvement but their participation). This is vital because they need to *own* the system.

In any particular case, the specific need should be articulated. It is important to distinguish between the need to obtain ISO 9000 registration from the benefits of having a formal quality system. If the need arises from customer pressure you still need to understand the benefits and how many customers require it and how much of your sales come from such customers. In some cases such pressure may come from as few as 20% of the customers but if they provide 80% of the sales then they have to be listened to. If the reverse is true, you have to look to the future and establish what the ratio will be in two years' time. It could change dramatically.

Regardless of customer pressure it is a good idea to assess the type of organization you have become and put these questions to the managers:

1. Are many employees unaware of the company's vision, values and mission?
2. Are we incapable of meeting all our obligations all of the time?
3. Are we capable of supplying nonconforming products or services?
4. Do we rely on verbal requests to get work done?
5. Do we hold on to our authority rather than delegate it to those on the scene of the action?
6. Do we often repeat the mistakes of the past?
7. Do we often have to repeat work because we misunderstood the requirements?
8. Do we expect our staff to know what to do without being informed?
9. Do we change our practices when we reorganize?
10. Do we consider our suppliers as vital to our success?
11. Do we expect our staff to learn to do their job without formal training?
12. Do we assign staff to jobs based on their availability rather than their capability?

13  Do we rely on managers telling staff how they want jobs carried out?

14  Do we often put the blame for our problems on to others rather than on to our own managerial competence?

15  Do we give priority to fixing problems rather than preventing their recurrence?

16  Do the deeds of the managers differ from what they say?

17  Do improvement initiatives fail to sustain the gains after a project is finished?

18  Would we have difficulty proving that we have done everything that we told our customers we would do?

19  Are we unable to quantify our true performance with respect to our key processes?

20  Are there any employees who are in doubt of the management's expectations?

If you obtain the answer 'yes' to more than a quarter of these questions then you have proven the need for a formal quality system. Even when there is some management support for such a system, many will hesitate and demand more information.

## Project feasibility

A review of the organisation should always be performed before starting the project, otherwise one may find oneself going in the wrong direction alone without any back-up support. All generals establish where they are and what they are up against before launching an offensive.

### Critical success factors

To ensure buy-in from all managers, establish the critical success factors (CSFs) for the business for it to achieve its mission (that is, the most important goals of the organization). They will be revealed from a knowledge of customer needs and expectations – not specific contracts but the factors that discriminate your organization from all the others of the same type or in the same industry sector. Such factors may include:

- Availability of service

- Reliability of product

- Capability of people, facilities, equipment, technology
- Capacity of plant
- Speed of response
- Image of company
- Value for money of product or service
- Accuracy and precision of results
- Integrity of relationships
- Security of information
- Confidentiality of information
- Customer care
- Prompt payment of invoices, wages

If this analysis has not been carried out before, then a complete and accurate list will not be produced at the first attempt. However, a partial list on which the management agree is better than nothing.

## Performance analysis

Before you establish the purpose of the system you need to find out where you are at present. Many companies start out by documenting what they do now and merely add the additional tasks needed to meet the minimum requirements of ISO 9001 or ISO 9002. This takes no account of their current performance and so after the system has been in operation for some months there is concern that it has not changed anything. In fact all it seems to have done is increased the amount of paperwork, created bottlenecks where none existed before and diverted scarce resources. Unless you establish your current position before you start you will have no means of measuring improvement as a result of installing a quality system.

A principle way of drawing attention to the change is to collect performance data that shows the current performance of the organization against the critical success factors such as:

1   Number of customer complaints

2   Number of breaches in security

3   Number of lost parcels

4   Defect rate, rework rate, repair rate (if a product supplier)

5   Quantity of unusable product produced

6   Response time to requests from a maintenance organization

7   Time to launch new products to market

8   Down time of process plant equipment due to failure

9   System availability (if a provider of public services)

10  Queuing or serving time (if a bank, building society, restaurant etc.)

Another approach is to compute the quality costs. That is, the costs incurred because failure is possible. Most business accounting systems do not measure the cost of doing the wrong things or doing the right things wrong. You may therefore have to estimate these costs. Whichever method you choose, it is essential to stick to the same cost categories throughout the programme, otherwise your conclusions will be invalid. It is the relative change that is important, not the absolute expenditure. Having identified where you are, you can then see what you want to change. Such costs comprise:

- **External failure costs** – the costs of warranty claims, repairs and other costs in correcting failures after delivery of product or service.

- **Internal failure costs** – the costs of correcting errors detected before product or service is delivered. This can include errors detected at any stage following release of the product or service design.

- **Appraisal costs** – the costs of detecting errors through inspection, test, review etc., including the cost of the equipment and facilities used.

- **Prevention costs** – the costs of preventing errors before work is performed, including planning and training costs.

It is important to present the results of the analysis in a form that will attract attention. Use the Pareto principle to separate the vital few from the trivial many.

It is also important to establish a baseline from which improvement can be measured after the project is complete. If you commence the project without doing so, it will be difficult to show how much improvement has been made and hence those that remain sceptical will not have ownership of the system.

## Preliminary review

A common practice before launching an ISO 9000 project is to carry out a preliminary review or gap analysis. This is a fancy phrase for finding out how far you are away from having a compliant quality system. If you already have a documented quality system that has been designed to meet other standards such as AQAP-1, Mil Q 95858A or Ford Q101, then such an approach may well be appropriate. It will tell you where the gaps are that have to be filled before you can apply for registration. However, if you have no quality system at all, a detailed account of where you fail to meet ISO 9001 or ISO 9002 is not helpful. The detail obscures some important messages. You only have to read the standard to realise how far away you are. A gap analysis could be like telling a non-athlete how far he or she has to go before winning an Olympic gold medal. What you need to know is the fundamental things to be done and you can usually determine this from a simple survey of the organization. In general, organizations do not have:

- Documented system of processes, resources and practices that control the business outputs

- Defined and documented responsibility and authority

- A formal means of identifying training needs

- Internal audit program

- System reviews

- A means for preventing the recurrence and occurrence of problems

- A culture that places commitment to quality as its basic belief

Many organizations do have some documented practices but they are usually based on organization functions and not formed into a system. At this stage, carry out a preliminary study of the existing documentation and determine the extent of use and its overall quality. A more detailed assessment will be conducted later on.

The gap analysis will address each clause of the applicable standard and identify the compliant and noncompliant areas. The analysis should be carried out against ISO 9004-1 or ISO 9004-2 rather than ISO 9001 or ISO 9002 as the later standards are purely for assessment purposes and not for building an effective quality system.

When doing such an analysis, it is important to think about what the management will do with the resultant report! If it is too detailed they will probably file it and in so doing miss any key messages you wanted to convey. Match the analysis to their needs:

- Do they want a detailed breakdown of conformities and nonconformities? If so, a clause by clause analysis is necessary.

- Do they want a general survey to highlight major strengths and weaknesses? If so, an element by element analysis is necessary, extending to no more than two pages.

- Do they want an expert's judgement on the culture?

## Cultural analysis

A cultural analysis will reveal much more about the probability of success than the other types of analysis. The *culture* is the values, beliefs and norms that permeate the organization and help to shape the behaviour of its members. It guides the organization in meeting its objectives, in working with one another and in dealing with customers, shareholders, suppliers and employees. There are many facets to an organization's culture and trying to characterize it is not easy. Here are some topics that can be used to reveal the culture.

---

### *Integrity*

- Are we honest in our dealings with customers, suppliers, employees and the public?

- Do we appear committed to quality by exhortations but then compromise quality when it suits our short-term aims?

- Is there integrity in our decisions or is profit the only driving force?

- Are we honest with the labour unions or do we tell them what they want to hear and then do as we please?

- Do we tell our bosses the truth or what they want to hear?

### Trust

- Do we trust those we appoint or do we look over their shoulders?
- Do we delegate authority to those we trust?
- Do we trust our suppliers to do what they say or do we check everything they deliver?
- Do we trust our customers to honour commitments?

### Ethics

- Does the management act the way it would have others acts and make no exceptions?
- Do we reward unethical behaviour?
- Do we put our own interests above our customers, suppliers and employees?
- Do we behave differently at work than at home?
- Do we take more out than we put in?
- Do we exploit our employees or treat them fairly?
- Are we more concerned with appearances than practices?

### Respect

- Do we respect the knowledge and capability of our staff?
- Do we respect the privacy and life style of our staff and resist the temptation to change it?
- Do we respect the decisions made by our superiors even when we may disagree with them?

### Change

- Are we proud of our traditions?

- Is our technology dated and inappropriate for today's needs or do we apply the best that will produce the desired results?

- Do we encourage change or fight it?

- Do we encourage innovation and creativity or stifle initiative?

- Do we pursue continual improvement or cherish the status quo?

- Are we of the 'If it ain't broke don't fix it' mentality or do we believe there is always a better way?

### Leadership

- Do we lead by example?

- Are we willing to take risks to achieve our goal?

- Are we willing to coach those less able than ourselves to achieve goals?

- Do we build teams with flexibility or are there demarcation, restrictive practices and work rules?

- Are we driven by objectives, by tasks or by fear?

### Consensus

- Are we autocratic in our decision making or do we aim for consensus?

### Communications

- Are staff kept in the dark? Does everyone know where we are going or is our course only known to the person at the helm?

- Do we encourage face-to-face communication?

- Do managers have an open-door attitude towards their staff?

- Do we listen to our customers, our suppliers and our employees?

- Is our communication both upwards as well as downwards?

- Do we autocratically enforce our unwritten policies upon our workforce or do we involve them when we form our policies?

### Reputation

- Do we have a reputation for quality, integrity and fair dealing?

- Do we have a reputation for being the best at what we do?

- Will we do anything unethical to protect our reputation?

- Do we care about our reputation?

### Participation

- Do we encourage staff to participate in decisions that will affect their future?

- Do we invite members of the public to view our facilities and participate in decisions that affect their environment?

## Planning

- Are we reactive or proactive?
- Do we anticipate events ahead of time or are we of the 'it might never happen' school?
- Do we plan ahead or wait for results before deciding what to do?
- Do we know first hand what causes the results or are we dependent on reports from our staff?

## Metrics

- Do we only measure our performance by the profit we make or do we also measure it by our effect on society, customers, suppliers?
- Are standards set and is performance measured or are we totally unaware of how the business is performing?

## Problems

- Are we constantly firefighting or are problems dealt with at the earliest opportunity?
- Do we consider firefighting as macho or do we value more highly those who prevent problems?

In assessing the culture you need to divide the results into Strengths, Weaknesses, Opportunities and Threats (SWOT). Some of the cultural traits will exhibit strengths that will drive the organization foward to success. Some will hold the organization back and act as barriers to success. Others will indicate a neutral position that may provide opportunities for positive change. And there will be those traits that threaten the very launch of the project.

Many quality initiatives have been undertaken over the last 15 years. Many have failed to deliver their promise largely due to management's lack of understanding of quality systems and of the relationship between quality and culture. A criticism of ISO 9000 is that it lacks the 'people focus' but TQM programmes that focus on people have also failed for the very same reason. The culture is created by the leaders in the company. Everyone follows the leader or is eventually retired. The leader shows the way and if this way is unethical then despite what is written in the policies and practices, the staff will behave according to the culture, not the written word. It is somewhat like the behaviour of motorists on a highway. When one is travelling at the speed limit and is passed by every car (and no police car is in sight!) one gains the courage to drive faster until every one is driving well over the limit. That is, until a police car suddenly appears and then there is mad panic to reduce speed before being spotted! If there is a rogue at the helm then ISO 9000 or any other standard is not going to be honoured. If on the other hand, there is a basically honest person at the helm who may be temporarily led astray by a few mavericks, all is not lost – change can be secured with the right treatment.

Determining culture is quite difficult as you can be misled by what some vociferous peole tell you. It is better to conduct a survey of employees that preserves the anonymity of the employee. The quesions above could be used to gather the data using a scale of the 'Strongly Agree to Strongly Disagree' format.

## Resources

How much should be put aside to pay for this project? Some organizations don't budget and then abandon the project when more urgent, short-term needs arise. Even those that do budget, often find themselves robbing the till to pay for the problems that arose too soon to be prevented by the new system. These situations cannot be avoided. The short term always comes before the long term. However, if a company were continually to operate like this it would never have any reserves for that rainy day.

The items to consider when estimating the costs are as follows:

- Management and staff labour
- Consultancy
- Training
- Set of ISO 9000 Standards
- Text books

- Documentation development software
- Communication software
- Data collection and analysis software
- Computer hardware
- Network architecture

If you really want to know how much it will cost use the following as a guide.

---

### Pre-launch

- Feasibility analysis – 1 man month
- Gap analysis – 2-10 days (includes writing the report)
- Management awareness seminars – 1-2 days
- Management decision making – 2 weeks

---

### System design and development

- Project Manager – full time
- Consultant – part time
- Steering Group meetings – one per month
- Process owners – one per key process part-time
- Support staff – as required
- Training courses – 10 days average training per team member
- Local workshops – 1 day per week for 6 months
- Software – £1000 upwards depending on number of licences
- Computer hardware – £1000 upwards depending on number of workstations

### System implementation

- Awareness seminars for staff – 1 day
- Local training workshops – 1-2 days for each employee

### System evaluation

- Auditor training – 3 days per auditor
- Management reviews – 1 per month leading to 1 per six months
- Internal audits – 2 per week for 3 months
- Pre-assessment – 1-3 days
- External audits – 1-5 days for 1-6 people
- Corrective action teams – as required

### System maintenance and improvement

- Management representative – full/part time
- System controller – full/part time
- Internal auditors – as required
- Management reviews – biannually
- Process owners – part time

For costs of consultants and registrars it is better to ask them to quote for specific tasks.

There is some truth in the belief: 'If I tell them the true costs they will not commit themselves but if I tell them only as much as they need to know, they may well commit to the project.' If you tell them it will cost £1m then obviously you're on to a loser, but if you give the costs for each phase at a time then the pill is easier to swallow! Cost the pre-launch and provide a rough estimate of Phase 1 with only the above outline for what to expect in the remaining phases. It will change anyway and you can't be accurate until you know:

a) The scope of the system

b) The number of documents to be produced

c) The number of staff that will participate

d) The consultancy support you will need

e) The certification costs

**Gaining commitment**

Commitment is doing what you say you will do. However, it depends on the doers knowing the right things to do. If a manager says he/she will award his/her staff with a 10% pay rise and does so then three months later the company goes bankrupt, the staff have been misled by a fool. A person who says they cannot commit to something because they don't have enough facts is being honest. One of the most common reasons why companies have trouble 12-18 months after the initial ISO 9000 registration is lack of commitment. They no longer do what they said they would do, but it is not because they are being deliberately dishonest (although a few may be in this category), it is because they did not fully understand what the whole thing was about.

The road to commitment is a seven-stage process, as shown in Table 3-1.

| Stage | Level | Meaning |
|---|---|---|
| 0 | Zero | I don't know anything about it |
| 1 | Awareness | I know what it is and why I should do it |
| 2 | Understanding | I know what I have to do and what I need to do it |
| 3 | Investment | I have the resources to do it and I am ready to deploy them |
| 4 | Intent | This is what I am going to do and how I am going to do it |
| 5 | Action | I have completed the first few actions and it has been successful |
| 6 | Commitment | I am now prepared do everything I said I would do |

Table 3-1 *The way towards commitment*

Resistance to change is often the result of ignorance, a lack of understanding of what is involved and a fear of being left to do it alone. At this stage in the project, you will not get commitment even to commence the planning phase until the executive understands what it is all about.

First there is *Awareness* and this can be gained through seminars, books, videos or inviting a reputable consultant along to give a presentation. It is often the case that the management will listen to an outsider rather than the change initiator as they may not respect the latter's views or may find it difficult to accept change initiated by someone of a lower grade.

There are two levels of *Understanding* to be grasped: firstly, understanding the concepts and principles; secondly, understanding the mechanics. The mechanics can also be divided into two levels. The first is how the project will be organized and the second is how the work will be carried out. Many people won't buy into a change until they know how it will affect them, often in some detail – so be ready for the inhibitors! Work on the innovators first – those who welcome change, whatever it is – then the conservatives who will change providing you show them the benefits, then leave the inhibitors to observe their peers and soon after they will follow or be left out on a limb[2].

## Unifying understanding

There needs to be a unified understanding of quality concepts amongst the management team. Some may perceive quality as striving for perfection, others may perceive it as inspection, as formalized quality departments or as increasing the overheads. It is highly likely that no one on the management team will regard new product development as quality improvement, market realignment as quality improvement or that you can have quality control in the design office! Some may perceive high grade as high quality, a Rolls Royce motor car to be superior quality to a family saloon.

It may also be likely that management takes the view that the workforce cannot be trusted and that independent inspection is therefore essential to achieve quality. Quality may be perceived only in terms of conformance to procedure, to drawing etc. and not to meeting customer needs. These perceptions need to be brought out into the open before you even decide on the kind of quality system you want. Your perception will determine what you include.

There are several standards that can assist with this understanding. However, beware – confusion may easily arise because they have been produced by different committees at different times and have not yet been harmonized. The standards you will need are:

---

[2] See *Managerial Breakthrough*, J. M. Juran (1964 – revised 1995).

- ISO 9000-1

- ISO 9001 (ISO 9002 is identical except for the exclusion of design control)

- ISO 9003 (if the quality of your product can be determined simply by final inspection before delivery)

- ISO 8402

- ISO 9004-1 (if you are a hardware provider)

- ISO 9004-2 (if you are a service provider)

- ISO 9000-3 (if you produce software)

**Climate**

When you have a picture of the culture, whether it be from a rigorous analysis or intuition, and you have the tentative agreement to the idea of change, the climate for change still needs to be right.

If the climate remains hostile to any such change after the nature of the project has been outlined, it is pointless proceeding. Either the approach taken has intimidated some of the managers or evidence supporting the case has not been convincingly presented. Many managers are preoccupied with control – maintaining the status quo. They have no time for improvement because they cannot leave the treadmill of control. They would like to change but cannot see their way to finding the time. The reader is recommended to study *Managerial Breakthrough* by Dr J M Juran. Amongst the many situations he describes may be found some possible solutions to the difficulties you may encounter.

Part of the problem may timing. The time for change has to be right. There is no sense in pushing for a change if the managers are already overloaded with changes. Pushing through a quality initiative when a big reorganization is underway or shortly to commence is fraught with danger from the outset. No one will take it seriously, let alone give any time to it. The organization has to be relatively stable to gain commitment from the managers. Even if the pressure is from customers, providing they are still buying from you, you may be able to delay the start of the project until the dust settles from the other changes.

## Consultancy support

### Selecting the consultant
Consultants are a resource. They have the advantage over permanent staff in that they can be turned on and off at will. But they can be expensive if not utilized properly. A consultant can offer expertise that you don't currently posses but you need to be sure of what you want before inviting a consultant to offer services.

Consultants can do several things:

a) Perform the gap analysis indicating how far you are away from your goal.

b) Assist you in convincing management of the need.

c) Provide awareness seminars to orient your executives.

d) Provide system development training to give your staff the skills to develop the system.

e) Guide you through the standards and the terminology.

f) Provide specimens of documentation.

g) Help you plan the project.

h) Help you choose the registrar.

i) Help you design the system and develop the documentation.

j) Help you through implementation.

k) Assist you with the project reviews, design reviews and management reviews.

l) Provide auditor training.

m) Conduct pre-assessment.

n) Advise on corrective actions.

Consultants can provide you with a set of manuals and procedures but they won't reflect your business and will more than likely confuse you rather than help you.

Consultants can operate as subcontractors, producing the documentation for you but at your risk! Most consultants will not offer this service unless they have the assistance of someone in the company. Too many organizations have chosen this approach, claiming they didn't have the time or the skills, and have been left with a set of documents that they find no incentive to refer to. There are a few areas where this approach works because the practices are new to the company (for example, auditing, management review, document control procedures). The consultant can draft the Policy Manual but it is recommended that you limit the assistance to drafting and get the managers themselves to complete it, otherwise it will always be considered as the work of the consultant and not their own.

Firstly, interview a few consultants. Don't rely on glossy brochures or even recommendations unless they are from a company in the same line of business.

- Spell out what you want the consultant to do.
- Establish what experience the consultant has of similar companies.
- Establish his/her track record.
- Establish the approach the consultant will employ.
- Establish what he/she will expect from you in terms of resources, time, activities etc.
- Treat the interview as if you were going to employ the person.
- Ask for a costed proposal and don't expect a firm, fixed price before the consultant has conducted an initial review of the organization.

If you approach a firm of consultants, you may receive a visit from a salesman. This person will probably not be the person assigned to the project. Always reserve the right to vet the consultant they have assigned. You need to feel you can work with the person. The consultant will be working with the middle managers so does he/she fit in with them? A person who is accustomed to working with the captains of industry may not be comfortable working with more junior employees and vice versa. The consultant also needs to have experience with companies of a similar size, although there is always a first time for everyone! It is largely a matter of attitude. Consultants emerging from large companies don't necessarily adapt well to small companies, but it depends on what they did in the large company. Many large companies are a consolidation of smaller businesses and have similar attributes to small companies. More importantly, the consultant needs to understand your business or have a wide experience that enables him/her to learn quickly and accurately from you about your business. Beware of consultants that are full of promises but don't deliver! A good test is to ask the consult to interpret certain clauses of ISO 9001 for your business. A consultant who says he/she doesn't know the answer but will find out may be better than one who tries to bluff through. Ask to see some examples of his/her work and form an opinion as to whether you believe he/she has the approach you like.

Many certification bodies are now taking a more practical approach so beware of the consultant who says that the auditor will not accept this or that! Look for someone who can give you options and help you to learn. Consultancy is about the transfer of skills and knowledge: How well do you consider your consultant does this? Does he/she keep everything close to his/her chest? Does the consultant have a personality and temperament that makes him/her approachable? You expect to learn from the consultant. You are the customer; therefore you can accept or reject advice – but before rejecting advice ensure the consultant enlightens you as to the consequences. Blindly accepting or rejecting advice either because you know no better or because you have not established a good rapport with your consultant is neither in your nor the consultant's interest. Consultants are human – or at least appear to be!

You also need to be aware of what the consultant will require of you. It is not a one-sided relationship. A consultant will only be effective if you provide the right information and give due consideration to each of his/her recommendations. Be prepared to explain your culture, your rules, your terminology, your organization and style of management. For a consultant to give appropriate advice there has to be a dialogue in which you talk freely about the things which characterize your organization. Should you hold back, believing it to be in the best interests of the company, you are not likely to receive the best advice. Consultants cannot give you sound advice unless you tell them the truth. Are you prepared for this?

You should not need to worry about confidentiality; your contract should protect you and most reputable consultants abide by a code of conduct that protects confidentiality. There are registers of consultants compiled through a screening process but you are no more certain of getting the consultant you require through this route than by going on a recommendation or selecting someone through an advertisement. Check out the credentials. If the consultant is a member of a professional institution or association you may gain confidence by checking this out. Most professional bodies have a code of conduct that covers confidentiality. If the consultant is also a registered auditor with an accredited registrar such as the IRCA or RAB[3], then confidentiality is also covered in the associated code of conduct.

## Setting objectives

Before the project can be launched, the objectives, purpose and scope of the system need to be defined. Everyone needs a clear understanding of why a quality system is being developed and what it is intended to achieve. At this stage, if you have decided you need a consultant, it would be prudent to include the consultant in these discussions.

---

[3] The IRCA is the International Register of Certificated Auditors in the UK; the RAB is the Registrar Accreditation Board in the USA.

## Target date

A target date for project completion needs to be set that is realistic and attainable with the resources that can be deployed. Don't set the start date earlier than when the necessary resources are available to commit to the project, otherwise there could be a delay and possibly abortive effort. When the staff do come on board they may want to change things around.

## How many systems?

It may seem an odd question but in large organizations it may not be practical to develop one system because the operations of each of its divisions may be quite different. One system can serve an organization in which all its operations contribute to a common purpose and mission and where there is one profit centre. If there are separate profit centres then separate quality systems may be necessary. It would be impractical for each department within a single profit centre to develop its own system because the departments would not be autonomous and would depend on the services of other departments to operate effectively. If the organization is divided geographically then separate quality systems for each region or country may be essential for market, cultural and language reasons. If divided by market segments, separate quality systems may be essential owing to the wide variation in customer requirements and product operating conditions. The aerospace sector for instance has quite different demands to the consumer-goods sector. In multinational corporations, there may be operational divisions with central marketing, personnel and research divisions. Owing to the disparate nature of these divisions, separate quality systems will be necessary and hence separate certification. Certification applies to a single quality system not a number of separate quality systems, unless they are integrated and function collectively as one system. As a guide, establish where the profit centres are and how business is contracted and subcontracted. The boundaries of the quality systems lie where money changes hands or costs are transferred.

## System scope

What you include and exclude is largely up to you. But if you intend going for certification there are some constraints. The quality system exists principally to ensure you supply conforming products and services to your customers. It follows, therefore, that your system should cover those parts of your business that contribute to satisfying customers. If you can separate parts of the business by customer sector, then you can pursue a phased implementation. If you can eliminate parts of your organization by specific products and services then you can limit the scope of the system. If, however, the separation cannot be made and attempting to do so would result in staff working to dual standards, then it is impractical to limit the scope of the system. You cannot say, for example, that the system will exclude those machines in the corner over there because they are costly to maintain, or we will exclude the site down the road and only cover the main site. If those machines and those sites contribute to the supply of the products and services that the other functions produce then they cannot be excluded.

The scope of the system is usually the scope of the business and therefore one needs to define what business you are in.

The scope statement needs to be succinct as it will appear in a directory should you pursue registration. The statement should define what products and services are to be covered by the quality system and where it will be applied by the organisation. It should be of the form:

> The (design, production, installation and servicing) of (products/services) for the ... (or) at the (plant designation, or market sector).

The following are some examples:

- The design, development, production, installation and servicing of mainframe computers and associated peripheral devices for military and civil applications

- The maintenance of sea-going vessels up to 200 tonnes

- The design and production of glassware for the domestic market

- The maintenance of air traffic control equipment for the London Air Traffic Control Centre

- The operation and maintenance of data processing facilities for use by the New York and New England Telephone Exchanges

### Defining system purpose

It is important that you know what it is that your system has to do – why it exists. A statement of the system's purpose should be developed that shows clearly why the organization desires a quality system and what it is to achieve in terms of the benefits to the organization.

The purpose statement should be clear enough for it to be used to determine the effectiveness of the system. There follow some examples:

- Benchmark current performance
- Bring operational documentation under formal control
- Create stability and minimize variance
- Eliminate complexity and reduce processing time

- Enable its installation at new locations without system redesign
- Enable the continuous improvement in operational effectiveness and efficiency
- Ensure products and services are delivered on time
- Ensure products and services satisfy customer requirements
- Focus attention on quality
- Harmonize practices on all sites to achieve consistency of purpose
- Improve efficiency
- Improve product and service quality
- Improve standards in those areas where performance is wanting
- Maintain those standards which we have been successful in achieving
- Provide a means for equipping staff with the capability to do their jobs effectively
- Provide a means for sustaining levels of performance
- Provide executive management with a high level of confidence
- Reduce customer complaints, internal defects etc.
- Reduce dependency on indispensable personnel
- Reduce dependency on informal practices
- Reduce operating costs
- Reduce quality costs

Whatever the reasons, the reasons need to be based on the facts revealed from the foregoing analysis. Producing a wish list will probably result in your being over ambitious! You don't need to go for zero defects at first or include all operations of the business. Quality improvement can be taken a step at a time. Some of the steps will be quite small ones, others may be quite radical and involve a major reorganization.

The measure of your current performance will indicate what needs to be improved and what needs to be maintained. There may well be some aspects of your performance that you want to remain at the current level because you are particularly good and competitive in that area. The quality system needs to sustain that performance even if in such areas you don't satisfy all the requirements of the standard. Formalizing the controls should not result in a reduced performance but in sustained performance. You will therefore come up with two types of objectives, some for control and others for improvement.

## Level of attention

Whilst the decision to pursue ISO 9000 registration will be an executive decision the attention it is given at each level in the organization will have a bearing on the degree of success attained. There are three primary organization levels[4] – the enterprise level, the business level and the operations level – and between each level there are barriers.

At the *enterprise level*, the executive management respond to the voice of ownership and are primarily concerned with profit, return on capital employed, market share etc. At the *business level*, the managers are concerned with products and services and hence respond to the voice of the customer. At the *operations level*, the middle managers, supervisors, operators etc. focus on processes that produce products and services and hence respond to the voice of the processes carried out within their own function.

In reality, these levels overlap, particularly in small organizations. The CEO of a small company will be involved at all three levels whereas in the large multinational, the CEO spends all of the time at the enterprise level, barely touching the business level, except when major deals with potential customers are being negotiated. Once the contract is won, the CEO of the multinational may confine his/her involvement to monitoring performance through metrics and goals.

The problem is that quality is a strategic issue that should involve the owners, as it delivers fiscal performance. If quality is low then fiscal performance will be low: if not in the short term, it will ultimately decline.

The typical focus for a quality system is at the operations level. ISO 9000 is seen as an initiative for work process improvement. The documentation is often developed at the work process level and focused on functions. Much of the effort is focused on the processes within the functions rather than across the functions and only involves the business level at the customer interface, as illustrated in Table 3-2.

| Organization Level | Principle Process Focus | Basic Team Structure | Performance Issue Focus | Typical Quality System Focus | Ideal Quality System Focus |
|---|---|---|---|---|---|
| Enterprise | Strategic | Cross-Business | Ownership | Market | Strategic |
| Business | Business | Cross-Functional | Customer | Administrative | Business Process |
| Operations | Work | Departmental | Process | Task Process | Task Process |

Table 3-2 *Attention levels*

---

[4] *Business Systems Engineering*, Gregory H Watson (Wiley, 1994).

With the typical ISO 9000 implementation, the senior management see it as a means to break into or stay in certain markets rather than as the means to sustain and increase the performance of the enterprise. Quality managers do not participate in the boardroom discussions because quality is perceived as a tactical issue, achieved only at the operational level – and of course this is not helped by the language of ISO 9000. Within the assessment standards there is little mention of strategic issues – quality policy being the only one. By interpreting organization structure as the way the enterprise functions rather than an organization chart, many more enterprise level issues would be brought into the quality system. By making quality a strategic issue, the quality system will become the means for growing the organization by better performance. The quality system should be designed to ensure all customers are satisfied with the organization's products and services. Consequently, if poor performance arises as a result of poor business performance, the quality system has failed to deliver. Getting this message across will ensure the quality system attracts the attention of the enterprise level.

At the business level ISO 9000 gets more attention but is often seen as a means of improving administration. The managers at this level instruct those below to implement ISO 9000 and appoint a manager to manage the project and act as their representative. By pushing the task downwards for implementation, the managers perceive it as a bottom up approach whereby actions are passed upwards for resolution but the running of the system rests with the line managers. Many ISO 9000 programs are functionally based and result in functional or departmental documentation that describes the processes used in the department. The training that is given is often targeted at the lower ranks (the task workers) and not at the managers. Much of the improvement effort is achieved in departmental teams rather than cross-functional teams. When cross-functional teams are formed they often comprise the task workers rather than the managers. By using the business process methodology to develop the quality system, the focus should change to the business level with cross-functional teams comprising the functional managers. These managers are the process owners not the task workers.

The customer is the primary focus of the business level and yet ISO 9000 is perceived to be weak on customer satisfaction. It is a weakness brought about by the language of the standard and not its intent. Its aim has always been to achieve customer satisfaction and requires the quality policy to be relevant to the customer needs and expectations. However, use of the phrase 'specified requirements' in ISO 9001 rather than 'customer requirements' has diverted attention away from customers. The managers at the business level have a clear understanding of how the business functions. They are concerned with business results and their measures of performance cut across departmental boundaries. A quality system designed around the business processes will involve all functions that contribute to customer satisfaction. Therefore, if customers are dissatisfied, the system has failed. Getting this message across will ensure the attention and participation of managers at the business level.

The attention of the staff at the operational level is essential but not until the upper levels have designed the system. Causing everyone to document what they do at the task level does not create a 'system', as will be discussed in Chapter 4. The system is an interconnection of processes that cause business results and prevent business failure. It follows therefore, that the task workers contribute to business success but as part of an integrated system, not as separate individuals. They are part of the chain of quality but linked through multifunctional processes. Many task workers don't see customers, don't know of the end customer requirements and so work to directives and specifications created by others who are used to translating and conveying customer requirements to the point of implementation. Task workers don't need to know the customer requirements provided that the other people in the chain have done their job properly. However, when designing a system that is intended to deliver products and services that consistently satisfy customers, the design has to be top down, not bottom up, so that customer needs are deployed correctly to the point of implementation.

## Appointing the Project Manager

No project should be started unless you intend to succeed so you clearly need to select someone who has the necessary skills, knowledge and personal attributes to lead the project towards success.

### Selection

The person who develops the quality system does not necessarily have to be the same as the person who will maintain it. It is not necessarily a long-term job. The Project Manager doesn't have to be a quality management expert but in selecting the right person look for someone who:

- Has the respect of the executive management
- Has the charisma to get the attention of both management and staff
- Possesses the attributes of a leader with the ability to encourage people to do the right things right
- Has the aptitude for planning and anticipating what is needed to reach the target
- Has the confidence to delegate and make others accountable for their actions
- Will motivate staff to produce results by persuasion rather than by fear
- Does not have other commitments that will cause unnecessary distractions
- Knows the business, the functions of each department etc.

- Knows the key personalities in the company and how to get results
- Represents quality in his/her actions and behaviour and hence sets an example to his/her colleagues
- Wants the job and regards it as of benefit to his/her career

Knowledge of quality management would obviously be an advantage but this can be acquired through training courses and literature. If you don't have a suitable employee, then hire one for the job. You don't have to place him/her on staff, they can be retained under contract. It can be a good way of selecting the right person.

**Role**

The role of Project Manager is simply to achieve the project objectives by using the agreed resources. Like any other managerial job, the Project Manager gets results through people – the people in this case being the representatives of each function that will be nominated to serve on the project.

It is not the job of the Project Manager to write the documentation. This is the role of the process owners. However, with a small company, resources will be limited and the project can easily become the brain-child of one person – the Project Manager. The Project Manager's primary responsibilities are to:

- Plan the project.
- Lead the development team.
- Organize work assignments.
- Determine priorities.
- Co-ordinate education and training of project team.
- Co-ordinate work and motivate those involved.
- Ensure the system is designed to meet the requirements.
- Produce/co-ordinate inputs to the Policy Manual[5].
- Report progress to management.
- Secure solutions to problems.
- Review and approve system documentation.
- Ensure release of approved documentation.

[5] Within this book the term *Policy Manual* is used rather than *Quality Manual* to indicate its true purpose.

- Drive system implementation and co-ordinate evaluation.
- Initiate contact with the certification body, review quotations and make recommendations to management.

## Training

If the Project Manager is not a quality professional or a practising Project Manager, some additional skills and knowledge may be needed, such as:

- Quality management principles and practice
- Quality system design, development and implementation
- Project management skills
- ISO 9000 standards
- Interpersonal skills
- Assertiveness skills
- Oral and written communication skills
- Personnel management skills
- Report writing skills
- Time management skills

## Common mistakes at the pre-launch stage

### Having the wrong reason

By far the most common cause of failure is having the wrong reason for creating a quality system in the first instance. Doing it to obtain a certificate will focus people on the goal for a short period and then when the goal is won, their attention turns away to something else. A quality system is not a goal, it is like a pet. A pet is for life not just for Christmas. A quality system is for life and not just for the certificate. And just like the pet, you have to feed your quality system, maintain it, care for it, exercise it, take it with you everywhere you go and it will repay you a thousand times. If your reason for embarking on this project is merely to obtain a certificate, then think again! This is the wrong reason. It is similar to a person undertaking a degree course just to get a degree. The right reason is to be educated in some area of knowledge, to a level that will prove to be beneficial in practice. Doing it merely to gain a degree will not be of benefit in the long term because you cannot do much with a certificate except put it on a wall. Sooner or later you will need to prove that you are educated. It is because some companies go for the certificate alone that ISO 9000 is not perceived as bringing any long-term benefits to product and service quality. The lack of immediate payback has led some companies to pursue TQM hoping this instead will give the desired results. There is no reason to abandon ISO 9000 just because the payback is not immediate. You have to use the quality system to bring about the desired improvements.

**The wrong purpose**

Using a quality system simply to document what you do will not yield the full benefits. In such a case the system is often viewed as being the documentation alone. If you think about it, when you document your practices you are taking a snapshot in time which soon becomes history. Every time you change a practice you should amend your documentation accordingly but this soon becomes a chore and lapses. If the system purpose was different it would be seen to cause things to happen. Ask yourselves what causes the results at present. It could be sheer hard work and dedication to duty. It could be a domineering boss who rules by fear. It could be pure chance. It could be that you have some very capable people. But can you rely on this alone? What would happen if these people were to leave, if the boss were to retire or move on, if your luck were to run out? The reason why electronic systems are more reliable today than they were some years ago is that the designers eliminated the causes of unreliability. They used components that had been proven to work when provided with the right signals. Electronic systems today do not work by chance, they do not depend on being hit with a hammer in the right place. In fact, strike them with a hammer and they fail! If the quality system is not going to be the sole means to cause the right things to happen then you are wasting your time. If you intend to continue to rely on the informal methods then sooner or later they will compromise the system and chronic failure is inevitable.

**Lack of resource**

Many false starts are caused by lack of resources. The CEO who assigns the task to a manager with other commitments soon finds the project has not progressed or, if it has, it is getting a luke-warm reception. One can proceed at a slow pace but it is not like a painting that you can pick up and put down at any time. With a painting you can see at a glance how far you have got. With a quality system project you can just as easily take two steps forward and three backwards if you do not keep up the momentum.

**Attention at the wrong level**

If the development of the quality system is given to the operational level alone, it is likely to flounder. Managers will not commit to ISO 9000 unless they are going to be measured against it. The task workers cannot design a quality system alone as they don't have the perspective necessary at their level. It needs a perspective from the highest level to see what is required.

A threat from the CEO to the Quality Management Representative (QMR) that his or her future rests in the balance will stimulate the QMR but no one else and such a program cannot be completed successfully without participation of all the managers. The QMR is the facilitator, not the instrument of creation. The CEO should make ISO 9000 registration by a given date an objective in every manager's departmental plan. It is insufficient to put the objective in the organization's business plan. It has to be passed down to all levels and appear at all levels as a goal to be achieved. Only in this way will it be taken seriously. Progress towards the goal should be measured at each level as a demonstration of commitment and the whole project will become a team effort of the company not just one department or one manager.

## Remote working

In proactive cultures, a Project Office is often set up and people seem happy to move their place of work. In reactive cultures it is difficult to get people to leave their comfort zone. They prefer to work in their office and communicate using e-mail, phone and fax. This reluctance to move can cause severe communication problems because remote workers do not form a good team. The team members have to get together, experience each other's enthusiasm and be able to bounce ideas off one another. Remote working may seem fine for the individual but is difficult for the Project Manager. If remote working cannot be avoided, you need a Project Manager who can manage at a distance, not one who is accustomed to having staff within arms reach; such a person will only talk to those closest and will leave the others out in the cold. Even if all the people are on one site, allowing them to stay permanently in their office creates a barrier to success.

## An unrealistic target

Demanding that registration will be gained in an unrealistic time puts pressure on the staff to cut corners (to avoid doing the right things and avoid doing things right because they haven't the time). Any task that the company has not done before is a journey into the unknown. You will not know of the obstacles you may encounter along the route and, hence, cannot estimate accurately the time each task in the project will take. You can only make a guess based on what others have achieved. Your consultant will obviously help, but he/she will not know of the barriers to success because they are unique to your culture. Resisting a change to the target date is like putting your head in the sand. It does not change the forces acting on the rest of your body. A target date that is set too far away in a reactive culture will not motivate the staff either, because they will believe they have plenty of time to deal with the tasks, only waking up to reality when certification is around the next corner with no time to learn from their mistakes. It is true that some people work well when under extreme pressure and others just crumble. You cannot expect everyone in your organization to be of the same temperament. You will inevitably have a mixture of personalities. Hence a time-scale that is only feasible for one type of person is doomed from the start.

## Summary

In this chapter the preparation needed to make a quality system project a success has been described. As the undertaking is not isolated to one person or department but involves everyone, it was shown how important it is to make adequate preparation before proceeding. When the need has been identified, it has to be proven to get the commitment of the senior management. Even at this stage it is unwise to go ahead without conducting a feasibility study that identifies the drivers and barriers to success. There is a need to quantify where you are now so that you can measure your success later. The culture was shown to be critical in planning the future strategy and identifying the areas requiring change. The costs need to be estimated and made abundantly clear to the management, as hiding them will cause resources to be diverted later when you need them most.

Some tips were provided on gaining commitment and hiring a consultant. The decisions you need to make (for example, the target date, the number of systems and the scope and purpose of the system) all have to be determined before you go ahead. Perhaps the most important decision, and yet the most difficult to implement, is the level of attention given to the quality system. Keep it at an operational level and it will not unite the processes that deliver the business outputs. It was shown that the project should be focused at three levels (the enterprise, business and operational levels) if it is to be of enduring value to the organization.

Selecting a Project Manager with the right combination of skills was shown to be important. Some common mistakes to avoid were covered, clarifying the importance of choosing the right reason, purpose, level of resource and attention, along with a realistic target to ensure project success.

## Pre-launch task list

1. Identify the reasons why you want a quality system.
2. Identify why you wish to certify your quality system.
3. Obtain copies of the relevant international standards.
4. Prepare a case for the quality system to present to your executive.
5. Obtain agreement to determine project feasibility before proceeding further.
6. Identify the critical success factors.
7. Collect data to indicate current performance against critical success factors.
8. Carry out a preliminary review against ISO 9004 to determine a baseline.
9. Perform a cultural analysis to identify the values, beliefs and norms that permeate the organization.
10. Carry out a SWOT on the results and judge the feasibility of project success.
11. Prepare a budget for the project obtaining estimated fees from consultants and registrars.
12. Commission executive awareness training to gain commitment.
13. Gain agreement on the terminology to be used.
14. Test the climate for change and decide whether or not to proceed.
15. Determine what you want the consultant to do and invite quotations.
16. Select and appoint the Project Manager.
17. Select the consultant.
18. Determine the project objectives.
19. Determine quality system purpose and scope.
20. Get buy-in at the right managerial level.

## Pre-launch questionnaire

1. Why do you want a formal quality system?
2. What do you expect from a formal quality system?
3. Why do you intend to seek registration of your quality system?
4. Who made the decision to create a formal quality system?
5. What are the project objectives?
6. What is the quality system intended to cover?
7. What functions of the organization will be excluded from the system and why?
8. What is your current level of performance relative to your critical success factors?
9. What factors will drive the project towards success?
10. What factors may act as barriers to a successful project?
11. What changes are being made in the organization at present?
12. What changes in the organization or in technology have been planned to be made during the duration of the project?
13. What resources have been allocated to the project?
14. How much time have the executive managers allocated to the project?
15. What action has been taken to gain the commitment of the executive managers?
16. Is the executive fully aware of its role in the project?
17. Are you clear on the division of work between the consultant and the company?
18. Is there a signed contract with your consultant?

## Chapter 4

# Managing the project

## Project management strategies

There are several strategies for managing an ISO 9000 project, each depending on the culture and size of the organization.

### The solo effort

In the small organization, someone will have been given the responsibility for achieving registration and it is likely that this would not be this person's day job. Time will have to be created for this additional responsibility amongst other priorities. The success of this strategy will depend on how much time and effort the Project Manager is able to put into it after dealing with other priorities. If the pressures of the day job are too much then the ISO project takes second place and never really has a chance of succeeding. If the Project Manager is not assertive, the project will drag on until the CEO decides what he/she wants. It is down to the CEO to reorganize the priorities. If the CEO wants registration badly enough then the earth will be moved to guarantee it! If markets and profits are more important, then ISO will always take a lower priority simply because it does not deliver short-term payback. Even with a dedicated Project Manager, little will be achieved. Unlike other projects, ISO 9000 is a project that should involve everyone; it can't be done by one person. It has to be a team effort, otherwise it will fail.

### The facilitation approach

In the larger organization, strategies vary. Some take the view that the 'Document what you do' approach applies to individuals, so the CEO orders all the department managers to get their staff to document what they do. Others take the view that departments or functions should document their practices and appoint a Project Manager to ensure they do this. This Project Manager is primarily a facilitator rather than a leader and success depends on the drive and motivation of this particular individual.

## Using cross-functional teams

If a business process approach is adopted the position of Project Manager becomes a full-time job for a short period because the co-ordination of cross-functional teams requires leadership. With this strategy, one can adopt a hands-on approach or a prescriptive approach, depending on the culture. Whilst some organizations appear innovative and advanced, they are often task-oriented. Tell some people what to do and they will do it. Give some people an objective and they have no idea how to proceed. The prescriptive approach may be necessary if the organization is task focused. If management by objectives is practised the prescriptive approach may impose too rigid constraints. However, awareness of a prescriptive approach will often help guide even the most innovative of persons.

## Applying ISO 9000 principles to the project

Project management techniques are, however, fairly standardized. BSI have recently published BS 6079 on Project Management and although intended for projects in the industrial sector, such as bridges, power stations, aircraft and spacecraft, the principles can equally be applied to other types of projects (and ISO 9000 is no exception). But there is no need to adopt other standards when one can use ISO 9001 as a guide to managing the project. The approach advocated in this book is based on applying ISO 9001 to the development of a quality system.

# Project planning

## A phased approach

The project should be divided into the following seven phases, definitions for which were given at the beginning of Chapter 3. The system maintenance and system improvement phases commence concurrently following the completion of the project, when the system is handed over for routine operation.

| | |
|---|---|
| **Phase 1** | Exploration (covered in Chapter 3) |
| **Phase 2** | Project feasibility (covered in Chapter 3) |
| **Phase 3** | Planning |
| **Phase 4** | System design |
| **Phase 5** | Documentation development |
| **Phase 6** | Implementation |
| **Phase 7** | Evaluation |
| **Phase 8** | System Maintenance Proving |
| **Phase 9** | System Improvement Proving |

Phases 1 and 2 may be complete before project start but there may be some tasks still to be carried out from Phase 2 so it is worth including them in the plan. Some of these phases will overlap, although no work should start until the project plan has been issued. Document development shouldn't commence until after the System Design Review (see later). Implementation can commence on release of the first documents and evaluation can commence following audit training and implementation of a set of documents making up a work process.

**Priorities**

The first priority is to obtain commitment by the management to provide the resources.

The second priority is the creation of an agreed system model of the business.

The third priority is the identification of process owners, after which design work can commence (but not without a plan).

Design always comes before documentation. There will probably be lots of existing documents. Some people will be eager to get started writing procedures. The Project Manager will have to resist the offers of help, for unless the project proceeds in an orderly manner abortive work may well result.

- You can't document procedures until you have defined what documents you need.
- You can't define the documents you need until you have defined the processes.
- You can't define the processes until you have modelled the business.
- You would be unwise to do any of these tasks until your staff have received appropriate training.

Before you start to produce the documentation, you need to put the documentation control procedures in place.

Before you create the document control procedures you need to design the documentation tools as part of the Quality Information System (QIS). This is addressed later in this chapter.

Before you assign tasks to people you need a plan (a means of allocating work such that everyone knows what they are required to do) and a means of determining progress.

Before giving out the work instructions you need to set up the Problem Reporting part of the QIS to capture concerns and queries and provide a common source of information. As sure as eggs are eggs, people will not do the right things right first time!

## Data requirements

The design of a quality system does result in lots of documents regardless of a desire to keep paperwork to a minimum. There are several documents or statements to be produced in order to record the organization's intentions. However, a document is merely a formal statement, which can be either a few lines or several pages, so don't be alarmed that there will be many document requirements.

A technique used in major projects is to specify data requirements for contract deliverables so that contractors have a clear idea of what outputs are required. This establishes expectations without the contractors having to go through a learning process (with the attendant risks that different results will be achieved). If we want consistency then the solution is to standardize.

Many ISO 9000 projects have resulted in a number of documents being produced that are based on the standard and don't vary from one organization to another. Nearly every ISO 9000 registered organization has a Contract Review Procedure, a Corrective and Preventive Action Procedure etc., regardless of what the organization does. The technique described in this book allows you to escape from this rigidity and produce documents that match the business whilst meeting the intent of the standard.

The *data requirement* is an instruction that specifies requirements for data that are required to be produced for a specified milestone. They enable the project to be driven by objectives and therefore serve to propel effort towards completion. If you don't divide the project into shorter targets, the project will flounder and never reach its destination on time. The data requirements provide a definitive requirement for people to work to stage by stage and so propel the project towards completion a step at a time. One could manage without data requirements and use task statements, but without some explanation as to the nature of the task, many tasks may result in wasted time.

An example of a data requirement is given in Figure 4.1 and a list of typical data requirements in Table 4-1. These data requirements are based on the system design methodology described in this book. You could use the same technique for any strategy you decide to adopt.

The beauty of the data requirement approach is that each data requirement has an identity and a deliverable that can be used in the project plan, and in reporting and reviews as a means of establishing progress. With vague instructions, one would never be sure where one was. With a definitive requirement, the excuse 'I didn't know what was required!' should not arise. Many people don't grasp the principles of ISO 9000 and quality systems for several months (if not years). People can work in a company for years without knowing why certain tasks are necessary – they do them because they are told to. You cannot wait until everyone understands quality systems and ISO 9000 so there will always be a level where you just tell them what to do.

| COMPANY X | DATA REQUIREMENT | DR01 |
|---|---|---|
| **Project:** | QUALITY SYSTEM DEVELOPMENT | |
| **Title:** | PROJECT PLAN | |

| Delivery Milestones | | |
|---|---|---|
| Draft | Intermediate | Final |
| Project Launch | Project Reviews | CDR |

**Input:** Project objective, ISO 9000

**Requirement:**

Prepare a Project Plan that identifies the tasks to be accomplished from project start to completion. The plan should cover the following:

1. Project quality policy
2. Project objectives
3. System requirement specification
4. Project strategy
5. Critical success factors and success criteria
6. Project milestones (phase descriptions and major decision points)
7. Project timeline
8. Project organization (chart and team responsibilities and authority)
9. Work allocation (major tasks, work packages, deliverables)
10. Resource provision in terms of office space, development tools, equipment
11. Subcontractor control plan (consultants, registrar, training providers)
12. Quality information system (strategy, tools and their development)
13. Communication plan (strategy, methods and tasks)
14. Education plan (strategy, eduction and training for executives to employees)
15. Evaluation plan (audits, design reviews and assessments)
16. Project reviews (strategy, project reviews and team reviews)

The plan shall be updated as the project progresses up to completion. It shall indicate the predicted and actual start and end dates for the tasks identified.

**Output:** Project Plan approved by Steering Group at a Project Review.

Figure 4.1 *Specimen data requirement*

# Managing the project

| DR No | Data Requirement Description | Draft | Inter | Final |
|---|---|---|---|---|
| 01 | Project Plan | Launch | | SDR |
| 02 | Action Item Log | | Weekly | FDR |
| 03 | Current Concerns Report | SDR | | |
| 04 | Vision, Values and Mission Statement | CDR | FDR | |
| 05 | Critical Success Factors | SDR | CDR | FDR |
| 06 | Current Performance Metrics | SDR | | CDR |
| 07 | Quality Policy | SDR | | CDR |
| 08 | Quality Objectives | SDR | CDR | FDR |
| 09 | Quality System Purpose and Scope Statements | SDR | CDR | FDR |
| 10 | Context Diagram | | | SDR |
| 11 | System Model | | | SDR |
| 12 | Process Models | SDR | | CDR |
| 13 | Register of Existing Documents | Launch | SDR | CDR |
| 14 | Glossary of Terms and Acronyms | SDR | CDR | FDR |
| 15 | Document Development Requirements | | CDR | |
| 16 | Register of Controlled Documents | SDR | CDR | FDR |
| 17 | Policy Manual | CDR | | FDR |
| 18 | Control Procedures | CDR | | FDR |
| 19 | Operating Procedures | CDR | | FDR |
| 20 | Blank Forms & Records | CDR | | FDR |
| 21 | Project Review Reports | | Monthly | FDR |
| 22 | System Evaluation Plan | SDR | CDR | FDR |
| 23 | System Evaluation Report | | | FDR |
| 24 | System Design Review | | SDR | |
| 25 | Critical Design Review | | CDR | |
| 26 | Final Design Review | | FDR | |

Table 4-1 *Data Requirement Matrix*

## Project plan

To establish a project plan you need four things: a knowledge of where you want to go (your objective), when you want to get there (the target date), when you are going to start (the start date) and how you intend to get there (the strategy). It's not easy, or usual, to get all four defined at the outset. Many organizations set a target date, assuming the start date is today. In fact, project start will vary depending on whether you are starting after or before the feasibility phase. Many start without a clear idea of the approach they intend to take.

It is not unusual for project planning to start with defining the timeline although a project plan is more than a timeline. This only tells you what and when. It does not tell you how. A plan is the provisions made to achieve an objective and clearly a timeline is only one of the provisions. With the start and end dates defined you can put two dates on your timeline. However, you need milestones en route so that you can pace your progress. Such milestones should reflect definitive stages of the project. Here are the three key milestones:

- **System Design Review** (SDR) – the stage at which system design is complete and documentation needs are identified.

- **Critical Design Review** (CDR) – the stage at which documentation has been produced and is ready for implementation (end of development).

- **Final Design Review** (FDR) – the stage at which implementation has been proven to be effective and ready for external assessment.

Allow three months between the CDR and the FDR so that sufficient objective evidence is obtained to demonstrate that the system is operating effectively before the external auditors arrive.

In addition to these there are six other milestones:

- Completion of feasibility phase

- Commencement of implementation phase

- Commencement of evaluation phase

- Establishing the post-development quality system management organization

- Proving system maintenance provisions

- Proving system improvement provisions

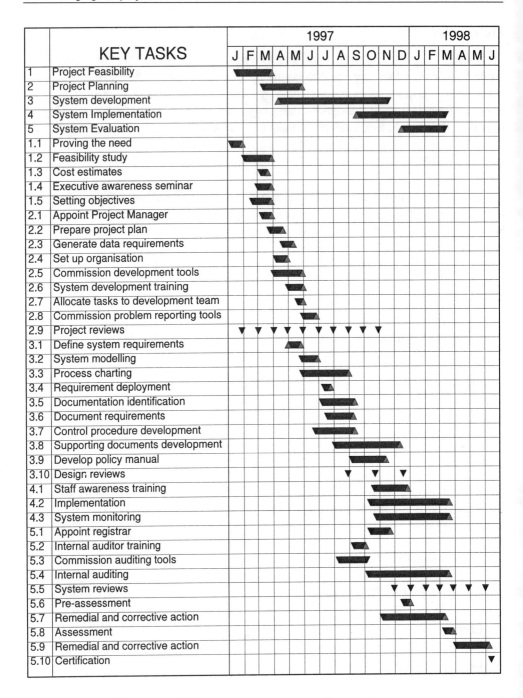

Figure 4.2 *Quality system development programme*

These will be sufficient for most programmes. Project timelines can be simple bar charts or detailed PERT charts, depending on the intended readership. The top management only need to see the progress towards the milestones, whereas the Project Manager will probably want to monitor each task of each phase. The phases overlap because one does not have to wait for all documents to be released before starting implementation.

Meetings such as the Steering Group meetings and Project Reviews can be added because they can be planned at regular intervals such as weekly, monthly or bi-monthly.

Any plan is more easily managed if phases are divided into tasks and tasks divided into activities. A typical set of tasks is included at the end of each chapter of this book. A typical project timeline is illustrated in Figure 4.2.

Often the timeline comes first and is sometimes considered the plan itself. It is but one part of the Project Plan as many important issues need to be resolved before you have a plan for achieving the project objective. The sample data requirement in Figure 4.1 above outlines the contents of a Project Plan. Other aspects are dealt with in this chapter.

## Organizing the project

The Project Manager needs a team, unless the organization is so small that the task is a one-person job. In an organization with several functions/departments, a team made up of representatives from each works best (so as to ensure full participation and ownership). If a quality department already exists then you have a choice. Either you assign the project to the quality department or you create a Project Team outside the quality department. The advantage of a separate Project Team is that it allows the existing quality department to function as normal without the distractions and without a reduction in resources. It also sends out a message that the project is multifunctional rather than the baby of one function. Place the project within the quality department and there may be conflict and pressure to maintain the status quo.

### The project organization

The project organization is a temporary organization built to last the duration of the project and is then disbanded. A typical organization chart is illustrated in Figure 4.3.

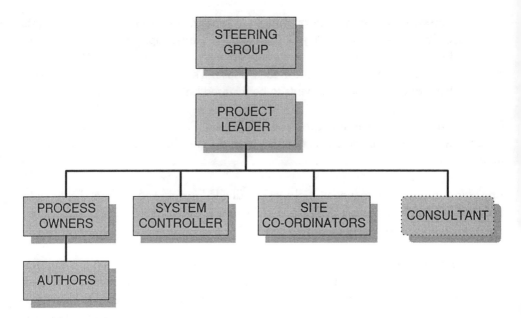

Figure 4.3 *Project organization chart*

## Project Steering Group

### Role
A project without a Steering Group is like a navy without an admiral. The captains may follow the course that is set but may easily get diverted by pressures from the crew. The Steering Group sets the course, provides the resources and checks periodically that the project is on course.

### Composition
The Steering Group should comprise the executive managers and be led by the CEO or a deputy. The Steering Group should own the project. Leaving the project entirely in the hands of the Project Manager when he/she does not directly employ everyone on his/her team may result in conflict with other managers. A common weakness of functional authority is that managers will have more than one master and will obviously serve the master that commands the greatest influence. A quality system project will be but one of the many projects within a company and so the priorities need to be set by top management. Only they can influence their line managers.

### Responsibilities
The Steering Group has one primary responsibility – the successful completion of the project. To make the Project Manager totally responsible is unrealistic as he/she will not have all resources under direct control. Other responsibilities are:

- **Demonstrating commitment**. It is important that members of top management show that they are serious about this project. It is not enough for them to sanction the resources. They will need to call the Project Manager to account for the progress and any problems encountered. They will also need to indicate the importance of the project to their own staff allocated to the project. Unless it is made clear that the project is of equal priority to other projects it will slip and waste resources.

- **Define the quality policy and objectives**. The quality policy has to be defined and documented by the executive management. This is a requirement of ISO 9000 that cannot be delegated.

- **Approve the Policy Manual**. This manual implements the quality policy and therefore top management must ensure that the operational policies are compatible with the quality policy. If they are not signatories to the Policy Manual they cannot complain when the policies do not achieve the results they expect.

- **Provide resources**. Top management holds the budget and allocates resources. A Project Manager cannot command resources that have not been allocated unless given a blank cheque. It is not unusual however, for some top managers to tell the Project Manager: 'Take what it needs to get it done!' Such cavalier attitudes are not helpful as other managers may feel their authority has been diminished. Such an approach rarely works in practice. Resources include staff, materials, equipment and perhaps contractors and consultants, none of which are inexpensive. The Project Manager will of course specify the resources required and carry responsibility for their utilization.

- **Project review**. The Steering Group needs to keep control through periodic reviews of progress. The Project Manager needs support, as do the rest of his/her team. It is not helpful to convene reviews only when there are problems, as this can be taken as a clear indication that top management is uninterested in the project. Even if the reviews are no more than 30 minutes a month, it is 30 minutes well spent.

## Process owners

When the system model has been developed, the identity of the key processes will be revealed. Each of these key processes should be assigned an *owner* – a person who has the authority to change the process should it be necessary to do so (either to meet the standard or to improve its performance). The process owners may be line managers or their nominated representatives. These people should form the quality system development team and be led by the Project Manager. The role of the process owner is to help determine the process models and identify the associated documentation (this is not a full-time job). The process owner may choose to produce the procedures or nominate suitable authors.

If there is more than one candidate for a position of process owner (simply because the same process is used on more than one site or in more than one department), the process owner would be the manager who is a level above these departments. It may be difficult to get this person to participate but it is his/her system that you are designing and he/she needs to be involved. The process owner can obviously nominate a representative but must participate in the decisions that agree the models and any changes thereafter.

The process owners' (or their representatives') responsibilities are as follows:

- Charting the processes
- Identifying the documentation requirements
- Appointing authors for the documents
- Co-ordinating the development of the documentation
- Providing policy statements for the Policy Manual
- Reviewing and approving process documentation
- Securing ownership
- Training staff in the use of the system
- Managing the implementation of the documented practices following their approval

## Authors

The authors should be the people who are most competent and able to produce the documents. They may operate at any level in the organization. Their authority is derived from their knowledge of the process or task to be documented. Often the people with the requisite knowledge don't have the documentation skills and ghost writers are needed. This obviously increases the cost and time required but you must refrain from eliminating the staff with the requisite knowledge just because they say they can't do it or are not available. They must be involved, as they will have the respect of their colleagues which is vital during implementation. If changes have to be made, there will undoubtedly be resistance and the best people to allay any fears are colleagues who work in the area concerned. If you still can't get their involvement then explain the consequences and report the situation to the Steering Group for their decision.

## Function representatives

Depending on the size of your organization you may also need representatives of each functional group to form a Process Development Team, unless the process owner is able to represent all functions associated with a particular process.

## System co-ordinator

Quality system development is initially paper-intensive while the policies and practices are documented. During this phase, draft documents need to be reviewed, changed, reviewed again, approved and issued. As the system is progressively implemented, policies and procedures will in all probability require change as knowledge of use is gained. You need to keep control of the documents throughout this turbulent period and this is a role for a quality system controller. For a small project, the Project Manager may perform the role but on larger projects you may need someone else. If it is decided to use computer-aided development tools someone with expertise in the chosen software application is vital to keep the project on course.

The co-ordinator will have the following responsibilities:

- Document generation and control software
- Ensuring network availability for document preparation and management tools
- Co-ordinating system change data
- Assisting the Project Manager
- Co-ordinator for training, meetings and communication

If you have more than one site at some distance away then a site co-ordinator may be needed (that is, someone to represent the Project Manager on a site and resolve the local problems).

## Consultant support

Should a consultant be appointed then the Project Manager is responsible for the appropriate utilization of consultancy time. A program of work should have been agreed before project launch but, as the work progresses, some alterations are inevitable. The consultant will have costed his intervention on the basis of your needs and what was found in the preliminary review. The extent of orientation required, both of the consultant and of your team, is difficult to predict until the consultant begins working with you. The cultural traits won't be recognized as drivers or barriers until tasks are attempted. This may alter the plan but not the overall costs, as adjustments can be made. Take time to educate the consultant in your business, your culture, beliefs and values.

A common difficulty is knowing what the consultant is doing and although any work will in all probability be in your interests, it might be misplaced. The consultant will work more quickly than you do simply because of his/her experience with such programmes. If the consultant has few clients then he/she will tend to put in more time. One who is heavily loaded may find difficulty in arranging convenient times for visits. The Project

Manager should keep a track of spend and deliverables. In fact it is good practice to monitor the consultant by deliverables, as this makes it easier for you to keep track and the consultant to meter the spend. You should both have an action plan to work to and limit spend to those items in the plan.

Another problem is relying on the consultant to force you into action. You may get into the situation where you only work on the project when the consultant is present. This is more likely when the project is being handled by one person who has other commitments. In these situations, you both get frustrated and progress is slow. It usually ends up with the consultant approaching the CEO to establish whether the company is serious about the project. It all comes back to commitment on both sides. If you don't keep the consultant busy then work for other clients will begin to take precedence and your project will be put on the back burner until you are ready to begin again. The consultant will attempt to keep you to the schedule but, if you resist too strongly, other clients will set the priorities.

If you don't listen to your consultant, reject or ignore advice frequently, the consultant may find it necessary to document every piece of advice and obtain a written response. More often, however, the consultant will have already deduced that the relationship is deteriorating and will request the contract be terminated. Good consultants should offer a range of options so that the client may choose one more fitting their culture and environment. Therefore, don't expect the consultant to give one answer to each problem. ISO 9000 is not black and white; there are lots of grey areas.

## Allocating work

Once the project plan has been agreed, tools need to be provided to allocate work, resolve concerns, monitor progress and direct remedial action. Microsoft Project or other project anagement tools can be used to lay out the plan and monitor progress. However, if the team is widespread, you will need a means of communicating requirements, capturing progress and dealing with problems. The data requirements define the quality of the outputs required but do not allocate work to individuals. One means of allocating work is to use a Statement of Work (SOW). In this document, which is usually no more than a single sheet, you can define the tasks each Project Team member is required to complete and the dates by which each task is to be completed.

There are, of course, other means such as a verbal request or a memo. It rather depends on your culture. If the norm is to issue verbal instructions you need to establish whether this will be effective. If the people concerned with the project have other duties (that is, they have not been allocated full time to the project), verbal instructions will, in all probability, not work. People forget and get diverted towards the manager who shouts the loudest. Memos, on the other hand, record the instructions but can often cause confusion unless they are organized: one memo saying one thing and another a few weeks later seeming to contradict the first. Above all, when you have a target to meet you cannot rely on informal methods for communication. Instructions need to be communicated through a controlled document and one such means is the SOW.

The SOW needs to address:

- Project name
- Work package (the parcel of work that is allocated)
- Work package manager (the person responsible for the work)
- Start date and end date
- Effort (if you are monitoring spend)
- Input requirements in terms of the data needed before the work can commence
- Requirements in terms of the tasks to be completed by reference to the data requirement where appropriate
- Output requirements in terms of the documents to be supplied

## Facilities

You will need a suitable room for people to meet for training and briefing sessions. Depending on the size of the Project Team, this could vary from a small office to a large conference room. There will be training sessions, so a room equipped with projection facilities is useful. When selecting a room, ensure it is fit for its purpose. A shed against the perimeter fence is not likely to impress the attendees and they will go away with the feeling that management are not committed! The other extreme is just as bad. Hotel conference facilities are splendid but a bit pricey – attendees may get the impression that management has money to burn! A room where you can leave charts on the wall that won't be disturbed is ideal. White boards and flip charts are all useful aids. If you equip the room with a computer and printer all the better as it allows results to be reviewed on the spot without delay. Often people will attend meetings unprepared and, unless you give them the results of the meeting immediately, by the time they leave and return to their desks they have already switched off.

The Project Team should be located together in a Project Office or 'Mission Control' so that work can be co-ordinated properly (see Chapter 3 for the rationale). Project teams do not behave as teams if their members are remote and have other distractions. Remote working only works when they are dedicated to the project and do nothing else.

## Quality information system

The *quality information system* comprises all the techniques used to collect, analyse and disseminate quality system information within the organization.

To begin with, the only materials you will need are paper and binders to contain the procedures. However, you may decide to use computers to produce the documentation or to have a paperless system using a computer network. You can prepare your procedures using a typewriter but desktop computers are now inexpensive and a good investment, as they save time and effort. You should obtain several software packages: a word processor for text production, a spreadsheet or database package for document control and a drawing package that can produce diagrams and flowcharts. The simpler the system, the less tools you will need. As you develop the procedures and standards you may find that you need other material and equipment to control the operating processes but these should be budgeted separately as part of the implementation programme. With either a paper-based system or a totally computerized solution, there are many tools that can be used to assist in the design, development, implementation, evaluation and improvement of your quality system. Here are some of the topics for which software tools are readily available:

- Auditing
- Meetings
- Calibrations
- Nonconforming material
- Capability studies
- Preventive action
- Communication
- Problem reporting
- Customer complaints
- Problem solving
- Document control
- Process analysis

- Document development
- Process mapping
- Failure modes analysis
- Sampling
- Gap analysis
- Statistical process control
- Gauge repeatability
- Supplier quality assurance
- Inspection
- Survey techniques
- Management
- Training

There are so many software products now available to 'help' ISO 9000 implementation that one is spoilt for choice[1]. In addition, there are the basic software packages such as word processors, databases, spreadsheets, flow charting and drawing packages. Basically, you need very few tools. Possibly as few as six tools that can share information to produce any number of outputs, from glossy documents to simple reports (as illustrated in Figure 4.4). Whilst you can obtain special software tools to perform each of the functions identified in the figure, many do the same job – the only difference is the subject. For instance, whether it's project, design or audits you are planning, a good planning tool will enable you do all three without buying separate tools. Whether they are customer complaints, nonconformities, potential problems or audit findings, a good problem-management tool will enable you to manage all this data, as the basic elements are the same. The problem is reported, the root cause established, remedial action taken

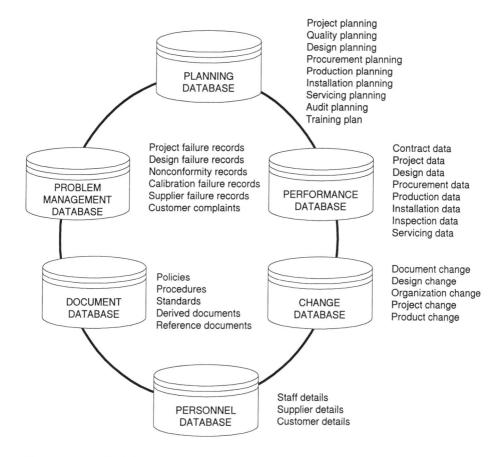

Figure 4.4 *Quality information system*

---

[1] ASC Quality Progress has an annual software directory which lists over 250 suppliers of QA/QC software.

and proposals for corrective action provided – all are common, regardless of the source of the problem. Whether it is the staff, suppliers or customers, details are often common such as name, address, descriptive data or some statistics. If you use common tools that are compatible with one another, data in one database can be used by an application in another database so that there is no redundancy and hence less chance of using erroneous data.

You do not have to locate the data under the headings shown in Figure 4.4. For example, the audit data can be stored in one place and the applications in another place so that the same tools are used to perform similar actions.

If you choose to develop your own database for the documentation it requires a lot of thought and effort to get it right. One weakness of any database is that once you have started to input documents, the freedom to change the database design becomes increasingly difficult, to a point where any change is very labour intensive. It is, therefore, vital that you fix your database design early in the project. If you choose a proprietary tool you need to evaluate it before you commence data input in any quantity, as it may not do what you want. Many software suppliers will give you demonstration disks but they are not sufficient for evaluation purposes. You need to try out the full package. More details of what you should look for with specific tools are given in the subsequent chapters.

It is advisable to prepare a QIS specification detailing your requirements and then evaluate the packages against these requirements. As you detect features you like, amend your requirements.

## Capturing existing practices

As soon as possible following project launch, make a list of all the documented procedures that are in use – the official ones and the unofficial ones. Establish their status and whether they reflect current practice. You may also have to search the memo files to establish management policy; without a formal system in existence, managers will probably have communicated their policies by memos, notices, minutes of meetings, standing orders, directives or any other similar method. Only collect the documented statements as the validity of any verbal instructions will be suspect. If possible, get management to review the information so as to discard any documents that have clearly become obsolete. You will probably find quite a lot that can be used as input data to the development process.

Present the results in a database or spreadsheet that can be easily amended and the data sorted and actions tracked. As the Controlled Document Register emerges you can indicate the status of documents as they move from the existing register to the controlled register. At the end of the project the existing document register should be empty because either the documents will have been converted to controlled documents or will have been destroyed or made obsolete.

## Education and training

The education and training of the Project Team is vital to its success. It should ensure you all work from the same hymn sheet and follow in the same direction, use the same terms and conventions and in effect work as a coherent team. However, this education and training needs to be delivered progressively. Give too much too soon, without the opportunity to practise, and people will forget what hey have learnt. There follows a list of education and training sessions that will be needed in the order that they should be delivered.

### *Executive briefing*
The executive briefing should be designed to give the executive managers an awareness of quality systems, ISO 9000 and the development process. The briefing should be delivered during Phase 1 in order to gain executive commitment. However, if this was not possible or not needed for this purpose, it should be delivered before staff receive any training, so that managers understand what they are doing and are more able to support their staff.

### *ISO 9000 awareness for implementers*
This training should be designed to give the Project Team an awareness of the quality system standards and provide understanding of the requirements that need to be met. The Project Manager could be sent on an external course and then deliver this course to the Project Team. Alternatively, the training could be delivered by the consultant.

It is not important that all staff receive this training. An in-depth course is better suited to the Project Team (that is, those who will do most of the work).

### *Quality system development*
Quality system development training should be designed to give the Project Team the skills and knowledge to develop the system. This training should be delivered to the Project Team including the process owners and function representatives. It is not effective if delivered to a wide audience.

### *Staff awareness briefings*
Staff awareness briefings should be designed to acquaint the staff with the new quality system, how it differs from current practice, and what their new responsibilities will be. The briefings can be delivered to all staff at one or more sessions or on an individual basis, tailored to the specific practices that will change.

## Quality system auditing

Quality system auditing training is usually a registered course delivered by external training providers and designed for those appointed as internal auditors. It is, however, equally suitable for the Project Team. Sometimes it can be beneficial for the team to receive this training before they commence documentation development as it provides them with an insight into what auditors will be looking for.

## Lead auditor training

Usually a registered course, lead auditor training is delivered by external training providers. Such courses are designed, primarily, for third and second party auditors but are equally applicable to the Project Team (particularly the Project Manager, designated management representative or Quality Manager). This training is beneficial prior to the commencing system evaluation.

## Specialist skills courses

Staff may need a range of skills to use the quality system in their environment. Such skills include:

- Problem-solving techniques

- Statistical process control skills

- Quality costing techniques

- Inspection, testing and calibration skills

- Supplier QA techniques

- Software metrics techniques

- Taguchi methods

- Benchmarking techniques

# Creating and sustaining the environment

The cultural analysis will have revealed the drivers and barriers to success. You have to run with the existing culture because a cultural change does not materialize overnight

---

[2] *Managerial Breakthrough*, J. M. Juran (1964 – revised 1995).

without a change in the leadership. Equipped with the same leaders, you need to adopt a strategy that will minimize the barriers and maximize the drivers.

Some people are innovators, some conservatives and some just plain inhibitors. Try to get the innovators on your team. Stay clear of the inhibitors as they can destroy the morale of the others. Work with the conservatives because they will work with you if you show them the 'why' and the 'how'. An unconvinced manager can sometimes have a change in attitude when working amongst peers who demonstrate enthusiasm. Make sure you conduct an awareness and alignment programme before you start so that everyone involved is prepared for the task ahead and so that any doubtful individuals are reassigned. There are three primary ways of creating the environment: commitment, communication and control.

## Commitment

### *Executive commitment*
Some executives are smitten by fashion. If they are not seen to be doing the same as their peers are doing then they must be doing something wrong. Some executives only follow their own instinct and no one else's – they do not commit to ISO 9000 unless they believe it will be profitable for the business. However, having an awareness of what ISO 9000 is and why it should be pursued is a long way from commitment. A CEO who stands in front of the assembled workforce to announce that the company is going for ISO 9000 is not necessarily someone who is committed to ISO 9000. The full six stages of commitment (see Chapter 3) have to be embraced, and action taken, before a CEO will be believed.

### *Staff commitment*
Staff will have different motives for what they do and in general will follow their leaders. If the leader is not committed to ISO 9000 then neither will their staff. A leader who sends out misleading signals can cause confusion in the minds of the staff. Saying one thing and appearing to do something else without explanation will unnerve staff. Even jokes about auditors can make staff cynical. However, even when a leader does show commitment, the staff may misunderstand the need and when faced with what may appear conflicting priorities, may act incorrectly. Staff need to feel that they can question a leader's requests – not to challenge them but to understand them. As Deming often stated, one has to drive out fear in an organization for it to realize its full potential. If staff are too frightened to question leaders, then the leaders are in danger of causing errors by default (that is, by not ensuring their staff understand the 'what', 'how' and 'why' of the work they do).

## Communications

If you find that certain things need to be changed before commencing the development programme, the changes should be made first before committing staff to the programme. The climate has to be right – other major changes need to be either far off or nearing completion before launching a quality system development programme. It is extremely important to gain commitment from top management before going ahead.

### Media

You need to agree on the communication programme: how the messages will be put over, when, by whom and to whom. If you have a company newsletter, then this vehicle could be used to announce the programme. You will need to replay the messages periodically throughout the programme, changing the emphasis where necessary. One problem you may face is the information gap which often follows the launch. You announce to everyone that you are going for ISO 9000, appoint the project leader and key players and then a long period of silence follows. By doing all the analysis and planning beforehand you can prevent this. Until you start, people have not been assigned and there may well be a gap whilst resources are assembled, people trained etc. Keep sending out positive signals. Management must keep demonstrating their commitment, their support and their interest.

### Computer network

If most of your staff work without the aid of computers then a computer network is unnecessary. However, if most staff use computers as an aid to carrying out their work and you have not considered using a network, the quality system development project provides a good opportunity to introduce network communications into your organization. A symbolic illustration of a computer network in given in Figure 4.5. Each workstation represents a location where a member of staff works or has access to the net-

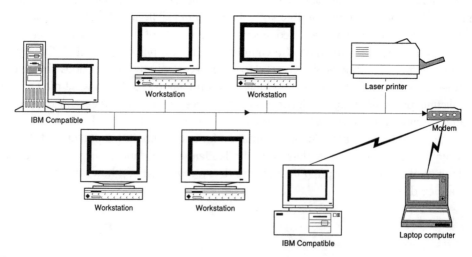

Figure 4.5 *Typical computer network*

work. These desktop computers are linked to a larger computer that acts as a server. The server is merely another PC with a high capacity data storage unit where the software applications and data are stored. One or more printers can be connected, thereby providing multiple printer options for users. A modem enables users at remote sites or on the move to access the server and use the system as though they were in the next office.

If you choose a computer-aided approach, electronic mail can quickly promulgate messages (with the added advantage that you will know who has and has not read the message). Networked workstations can also display messages at 'switch on' although staff will get desensitized to these after a while. However, beware – messaging is not document control. There is a significant difference between 'groupware' and document management tools (see Chapter 7).

## Control

The Project Manager needs to be in control of the project. This means that the project has to be agreed by those affected as well as by those providing the resources. Thereafter, the Project Manager should hold reviews at defined periods to monitor progress, identify problems and seek their resolution. There are two methods that can be employed. One is the *project review* and the other the *design review*.

### *Project reviews*
Project reviews should be held with the Project Team, including the consultant, to review progress to date. This is where the data requirements aid tracking. If every planned action has a data requirement then the task is predefined and verification of completion made easy. The project review is simply a review of progress if you intend to review the system design separately. If you combine the reviews then at some stage the design needs to be agreed. Otherwise, the design will evolve without any defined end point and with succeeding tasks commencing before design has been completed.

### *Design reviews*
Design reviews should be held to agree the design of the system. A quality system is designed first, then documented. As the system is a means for causing the right things to be done, the processes that do this need to be designed. Although these processes will generally reflect the way the organization operates, tasks need to follow in a logical order, with the right inputs and outputs. The processes need to reflect best practice and, therefore, changes may be needed to achieve this. The design is the way these processes interface with one another in a continuous chain which delivers the products and services. There are three levels of design: completion of the system design; completion of documentation development; and completion of system evaluation, proving that the system is effective. (This equates to the preliminary design, detail design and design verification and validation stages of developing a product.)

## — System Design Review (SDR) —

The System Design Review should be held at the stage at which system design is complete and documentation needs have been identified. A list of deliverables for the System Design Review is given in the box. These represent all the key decisions that have to be made in order to proceed with documentation development.

The system design should be reviewed by the Steering Group and the Project Team should present the design data as defined in the data requirement. The result should be an agreement to proceed to the next stage of development.

The key aspects to be addressed at the SDR are as follows:

> **SYSTEM DESIGN REVIEW DELIVERABLES**
>
> 1. Project Plan – Final
> 2. Glossary of Terms – Draft
> 3. System Model – Final
> 4. Context Diagram – Final
> 5. Register of Existing Documents – Intermediate
> 6. Critical Success Factors – Draft
> 7. Quality Objectives – Draft
> 8. Process Definitions – Draft
> 9. Documentation Register – Draft
> 10. Policy Manual – Draft

1. Does the system model accurately reflect the business?

2. Do the key processes accurately reflect how we control product and information as it passes between the various functions of the business?

3. Are the definitions of the processes consistent with our business objectives?

4. Have all external and internal interfaces been accounted for?

5. Do the models show that the controls are compliant with the governing standards?

6. Are the models consistent and coherent?

7. Are the critical success factors aligned to the mission statement?

8. Are the changes to existing practices consistent with the standard and our long-term aims?

9. Have the required key documents been identified?

### — Critical Design Review (CDR) —

The next stage is when the documentation has been completed. A list of the deliverables for the CDR are given in the box. They represent all the key decisions that are needed in order to proceed with full implementation. Implementation will have commenced before completion of the documents but not until the control procedures are drafted. It is important to vet the interfaces before operating procedures are produced. If this is not possible, there may be some redesign necessary. The objective is to try to avoid too many design iterations. If the documentation is being developed by a single person, such controls are unnecessary. When you have a team developing documents, it is vital that coherence is maintained, otherwise rewrites are inevitable. Key aspects to be addressed at the CDR are:

> **CRITICAL DESIGN REVIEW DELIVERABLES**
>
> 1. Quality Policy – Final
> 2. Quality Objectives – Intermediate
> 3. Glossary of Terms – Intermediate
> 4. System Model – Revised
> 5. Register of Existing Documents – Final
> 6. Critical Success Factors – Intermediate
> 7. Process Definitions – Intermediate
> 8. Documentation Register – Intermediate
> 9. Policy Manual – Draft
> 10. Document Development Requirements – Final
> 11. Control Procedures – Draft
> 12. Operating Procedures – Draft
> 13. Blank Forms/Records – Draft
> 14. System Evaluation Plan – Draft

1. Have problems detected at the SDR been resolved?
2. Has integrity of the process models been maintained?
3. Do all outputs from documents serve as inputs to other documents?
4. Is there a coherent trail from one document to another?
5. Has the use of forms been minimized?
6. Are common routines covered by common procedures?
7. Do the procedures cater for all the optional paths we need to maintain flexibility?
8. Are there any gaps in the description of the processes?
9. Have standards been specified for decisions that affect quality?
10. Are the specified responsibilities clear and unambiguous?
11. Do the procedures reflect best practice?
12. Have the necessary purchases been made to implement the system?
13. Have the necessary human resources been allocated to implement the system?
14. Has the necessary training been planned?

## — Final Design Review (FDR) —

The final stage is where the system is ready for external evaluation (the validation by third parties). A list of the deliverables for the FDR is given in the box. They represent the key decisions in confirming the readiness of the system for assessment. The key aspects to be addressed at the FDR are as follows:

1 Have problems detected at the SDR and CDR been resolved?

2 Have all planned documents been developed?

3 Have audits been conducted to verify that the documented practices are in fact being followed?

> **FINAL DESIGN REVIEW DELIVERABLES**
>
> 1 Quality Objectives- Final
> 2 Glossary of Terms – Final
> 3 Critical Success Factors – Final
> 4 Process Definitions – Final
> 5 Documentation Register – Final
> 6 Policy Manual – Final
> 7 Control Procedures – Final
> 8 Operating Procedures – Final
> 9 Blank Forms/Records – Final
> 10 System Evaluation Report – Final

4 Is there evidence to show that we are doing what we say we will do?

5 Are the practices compliant with the requirements of the governing standard?

6 Have all practices been tested?

7 Is there evidence to show that the system is being improved through corrective action programmes?

8 Have the necessary resources been allocated to maintain the system?

### Staying on course

Staying on course is not easy for a project such as this because the Project Team members are probably not devoted full time to the project. There may be staff changes, emergencies, reorganizations and other distractions. Keeping the project on course requires dedication and commitment. Progress should be addressed at each Steering Group meeting and each meeting should be held as planned. The project should be among the objectives of each of the executives and the Project Manager should be frequently held accountable for progress. The momentum needs to be sustained. Issuing bulletins of achievement can help to do this. Success breeds success so, when the first process audit has been successfully completed, announce it to the troops. Don't move the target date unless it is hopeless. Keep the pressure on both the management and the staff!

Try not to change the Project Manager or the consultant unless, of course, the Project Manager or the consultant is the cause of the delay. Continuity is essential to success. No one likes a change in direction so by plotting the correct course at the outset, and staying with the same team, you are more likely to stay on course than if you keep chopping and changing in pursuit of the right mix of personalities. There will be problems but they can be overcome with good managerial skills and the application of common sense.

*Problem database*
Before anyone is given a task, a means of handling problems needs to be established. If you wait until the problems arise you will end up firefighting with no clear idea of who reported what problem, what was done about it and whether it was resolved. A simple database that uses forms for people to submit information on a problem to the Project Office and receive a response is all that is needed. The Problem Reports should be indexed and categorized as to the type of problem so that you can quickly see where the most confusion arises. Status information also needs to be provided so that you can quickly determine which problems have not been addressed, which are being worked on and which have been closed. It is a most valuable tool for any project as it puts the data in a place where problems can be tracked. It also provides objective evidence of the principle concerns to management should they doubt your report to the Steering Group.

## Organizing the assessment

### Selecting the certification body

The certification body, or registrar, is the organization that carries out the third-party assessment of your quality system against one of the quality system standards. When selecting an appropriate registrar there are several factors to be considered.

*Accreditation*
There are two types of registrar – accredited and non-accredited. The accredited registrars have been assessed by a national accreditation agency such as UKAS, RAB or RvA and found to meet the requirements of EN45012 for the certification of quality systems. This standard, although European, is recognized world-wide by other accreditation agencies and registrars. Eventually an international standard will emerge.

The non-accredited registrars are offering similar services to their accredited cousins[3] but without the assurance of quality of service that accreditation bestows. The accredited registrars are accredited for performing audits in specified industry sectors, hence some registrars are accredited for sectors for which others are not. A registrar that is not accred-

---

[3] *Quality World* (December 1996) and *Business Standards* (May 1997).

ited for your industry sector but accredited for others is in a different class to those registrars that have no accreditation at all!

The non-accredited registrars are cheaper and offer complete packages. Beware! They look and sound very plausible. The certificates look very similar and you can be taken in by the sales patter. If you want to use the certificate to trade, then you will be asked by your customer to identify the registrar and registration number. A simple investigation will reveal the certificate to be worthless. Not only have you wasted your money but you have lost some credibility as well!

However, if accreditation is not important to you and you merely want an independent assessment then go ahead. There is no law preventing these organizations from selling you their services. You get what you pay for: no more, no less.

### Requests for quotation

The first stage, therefore, is to choose a number of registrars that are accredited for performing quality system certification audits in your industry sector. A directory of accredited registrars can be obtained from your national standards body. Alternatively, a survey of 'quality' magazines such as *Quality World* and *Quality Progress* will reveal details of many accredited registrars. The non-accredited registrars do not generally advertise in the 'quality' press. You may find that their approach is to put a leaflet through your door or make sales calls or visits.

Before requesting the quotation you should have defined the purpose and scope of your quality system so that the registrar can establish whether he or she has the capability to do the assessment. Request quotations from at least three registrars. The registrar will, in turn, send you a questionnaire to complete with details about your company. The reason for the detailed questions is that they need to establish the size and complexity of the company in order to calculate the number of audit days.

### Evaluating the quotation

The problem with quotations is that they are easy to judge on price but difficult to judge on quality. It would therefore be prudent to supply a questionnaire that will aid your judgement of the quality of the services offered. Questions addressed to the registrar might include:

1. What experience have the auditors in our business?

2. Which organizations of our type have you registered?

3. What continuity will be maintained between auditors chosen for the initial audit and the surveillance visits?

4   What criteria will be employed to judge whether the quality system meets the standard?

5   What are the conditions for issuing a certificate?

6   How soon after the assessment will we receive the written report and what will it contain?

7   What approach will the auditor take to the audit – will he/she be checking element by element or process by process or department by department?

8   Will the auditor wish to visit every site?

9   How much notice do we have to give for the initial assessment?

10  When will the auditor want to review the documentation?

11  Which documents will the auditor wish to examine in the documentation review?

12  Will the auditor want to review the documentation on or off site?

13  How will the results of the documentation review be conveyed?

14  Will the auditor require objective evidence of compliance with every requirement at the time of the audit or can some requirements be checked at a later stage?

15  Will you permit us to vet any new auditors before their assignment?

16  What guidance will be provided to help us improve the system?

17  Should we wish to lodge a complaint about the audit, what is the procedure?

Some of these questions may already be addressed by the literature you are sent, but a discussion to clarify these issues is often useful for both parties.

## Common mistakes in project management

### Not steering the project

Once the project is underway and the management have confidence in the Project Manager, there is often a desire to leave well alone and wait for the Project Manager to report exceptions. There will also be other priorities of the business that may distract management attention. Depending on the culture, the Project Manager may not wish to report bad news. No one likes bad news; therefore it gets suppressed unless there is a culture in which staff feel free and uninhibited about reporting actual or potential problems. Failing to steer the project may result in surprises when it is too late to divert resources to fix the problems. Failing to steer the project also sends signals to the other managers that the project is not important.

## Not understanding process management

The most common approach taken to an ISO 9000 project is a departmental or functional approach where each manager works within the confines of his/her own department. With the *process management approach*, process owners take charge of the development and form cross-functional teams to develop a coherent system. If the process owners do not embrace process management properly and retain their functional mind set, they will not cause process-based documentation to be produced. Whilst the control procedures may reflect a process approach, the supporting documentation may well remain functional. The net result may be a fragmented system with isolated pockets of resistance, hanging on to departmental traditions. This confusion may result in several documentation systems where one would be more effective. The confusion may also result in several quality systems, especially in those departments that have little interface with others.

## Distractions

Distractions are the enemy of quality system development. They divert resources, focus priority away from the project and generally demotivate the Project Team members who are only trying their best to meet their objectives. Common distractions are reorganization, takeovers, mergers, new technology – in fact any change that affects the staff. Launching a quality system project when the working environment is changing is not a good omen for success. Sometimes the management realizes it is impossible to execute the project with so much change and suspend work until the seas are calmer. This also has a demotivating effect and when staff are reassigned to other duties the chance of getting them back diminishes. It is better not to start in the first place if major changes are anticipated. If you can't allow adequate time then the management needs to be prepared for the consequences.

The initial assessment by the third-party auditors is but a milestone. It is not the end of the road, in fact it is only the beginning. Depending on your choice of certification body, you may be fortunate to have selected one that is not thorough on the first assessment. Although you have 90 days to correct the nonconformities, it could be costly if you have major nonconformities which necessitate a re-assessment.

## Weak leadership

The success of any project depends upon the leadership qualities of the Project Manager. If the person selected is weak and allows other managers to browbeat him/her, then the company will get the quality system it deserves. There is no point in blaming the consultant as the consultant is not at the helm driving the ship. If the Project Team comprises strong personalities who are able to influence management then all is not lost. A weak leader may be reluctant to issue instructions to others. A strong leader lets everyone know where they are going and keeps an eye on those who might stray, bringing them back on board to serve the team. A strong leader builds a team in which there is enthusiasm for the project – where each team member makes the right contribution and works in harmony with the others to achieve the common goal.

---

### Management task list

1. Meet with selected consultant.
2. Agree project strategy.
3. Deliver Executive Briefing if not carried out previously.
4. Get commitment to the project from executive management.
5. Appoint Project Team members.
6. Arrange facilities such as office space, training rooms, computers etc.
7. Determine project milestones.
8. Produce the project plan.
9. Select or produce data requirements for project deliverables.
10. Prepare Statements of Work for team members.
11. Establish a problem management database to capture concerns.
12. Convene first Steering Group meeting to agree plan and Project Team members.
13. Commission Project Team training.

14. Initiate communication media.

15. Set up documentation tools.

16. Co-ordinate system design activities.

17. Capture existing documentation.

18. Convene monthly Steering Group meetings.

19. Convene weekly project reviews with Project Team.

20. Monitor progress on data for System Design Review.

21. Conduct System Design Review.

22. Deliver documentation training.

23. Monitor progress on data for Critical Design Review.

24. Conduct Critical Design Review.

25. Deliver staff awareness briefings.

26. Request quotations from suitable certification bodies.

27. Select certification body.

28. Commission auditor training.

29. Monitor progress on data for Final Design Review.

30. Conduct Final Design Review.

31. Commission pre-assessment.

32. Resolve nonconformities.

33. Commission third-party assessment.

34. Establish the organization to manage the system after the assessment (see Chapter 9).

## Management questionnaire

1. Have you got the commitment of the executive management for the project?
2. Do all the executive managers know what they have to do?
3. Has agreement been reached on the resources needed?
4. Have the key milestones been agreed?
5. Has the project plan been agreed by the executive management?
6. What induction have the executive managers received?
7. Have all existing documents been registered?
8. Have dates been set for Steering Group meetings?
9. Have the process owners been nominated and trained?
10. Does everyone on the Project Team know what is expected of them?
11. Do you know what the consultant will provide for the company?
12. Do you know what tasks the company have to do?
13. Have you identified any inhibitors and ensured they are not on your team?
14. Have you agreed how and when communication with the staff will be handled?
15. Have you requested quotations from at least three certification bodies?
16. Have you interviewed the prospective auditors or sent them a questionnaire?
17. Are you continuing to conduct progress meetings and design reviews?
18. Are steering group meetings being held on schedule?
19. Are problems/solutions passed to the executive being promptly resolved?
20. Will you make the target date?

## Chapter 5

# System design

## The nature of systems

In Chapter 1, Alan discovered that quality systems are not a collection of procedures but an interconnection of processes that deliver desired outputs. These outputs are the products and services supplied to customers. Therefore, the system has to cause outputs of the required quality and prevent outputs of poor quality. To cause anything to happen there has to be a chain reaction. The chain in this case extends from the customer requirements through all the processes that convert these requirements into products and services and out to the satisfied customer. These processes cross each function or department in the organization that contributes to achieving customer requirements.

What makes a collection of components into a system is the way they are connected and energized. If we were designing a projection system for projecting images onto a screen we would need a source of power, a light source, a lens to focus the light and a platform on which to place the image we wanted to project. Individually these components do not make a system but, when we assemble them into the right order and make the necessary connections, we have a projection system. Some of the components themselves are a collection of other components and we can refer to these as subsystems if they fulfil a unique function within the system. An example of a subsystem would be the power supply. Disconnect the power source and the system fails. Damage the glass on which the image sits and the system fails. Fail to fit the lens the right way round and the system fails. Only fit one light bulb and when it fails the reliability of the system is reduced. Design the system so that you can't get at the bulbs without a screwdriver and the system fails to fulfil maintainability needs. Equip the instrument with too low a light source and the image isn't bright enough to be seen clearly. Design the lens head as a fixed structure and it requires greater storage space when not in use.

In an air-conditioning system, as another example, there are many components arranged in a certain order. There are the power source, the heat exchanger, the room thermostat and the cables. Cut the cable to the sensor and no matter what you do you can't adjust the conditions in the room. Fit the wrong heat exchanger, and the demand you make when setting the thermostat won't be met. In this case there has to be a mechanical feedback loop for the system to function. In the case of the projection system, the feedback was provided by the user.

When we design systems, we start not with a set of components and see what we can make from them but with an objective and proceed to produce a diagram showing the primary functions and how they are interconnected to achieve the objective. For any system to work, the components have to be of the right type and connected in the right sequence and supplied with the right energy. There is also a need for sensors to detect demand, sensors to monitor system performance and functions to adjust performance so that demand is met every time. These examples focus on *closed systems* – systems that contain all elements that enable control of the outputs within the system. The cause and effect relationships are within the control of the system. One might argue that market forces, the economy and the supply of labour and materials are outside our control but we do have in our power the ability to regulate what we do as a result of changes in these factors. Many companies cease to trade in a depressed economy but there are also others that continue to sail through all types of weather because they have the right formula for success.

Let us take an example of an *open system*. An education system can be perceived as an open system as there is no central control over all the elements. In such a system there are many different types of component:

- Government legislation
- Academic standards such as national standards, syllabus, curriculum
- Buildings such as schools, colleges, universities
- Equipment such as laboratory instruments, computers, sports
- People such as teachers, parents, students, examiners, employers
- Extra-curricular activities – supervised and unsupervised
- Materials such as text books, examination papers, teaching aids

Each represents a subsystem and some are not wholly dedicated to the education system (such as the buildings, materials equipment and employers). However if we treat the whole as a system with uneducated people being the inputs then the output from such a system should be educated people that benefit society. If the system turns out people who are of no use to employers or who vandalize our society then the system has failed. If employers were to have to seek staff from overseas then clearly our system would have failed. If the environment within the educational establishments is such that it deters teachers entering the profession, then the system has failed. The problem with this type of system is that there is no single owner. It can only be society that owns the system. We make it what it is; we get what we pay for. The cause and effect relationships are not as well defined as with closed systems. Other large systems such as health care and transport also have this problem and can only be dealt with at the subsystem level. If we can control each of the subsystems then we may cause the system to produce the desired outcomes. One system that is certainly out of control is our ecosystem because there are too many factors that influence it. Bringing the ecosystem under control requires international co-operation on a huge scale. Many components are designed for one system when in fact they are used in other systems. The automobile was designed as part of the transport system, but is also a component in the ecosystem, and in fact a component that is not working at its optimum for both systems to work effectively. An analogy within business is the organizational component such as a department. If we treat each department as a system then its operations can be optimized. Treat it as a subsystem of a larger system and we find that when each department is at its optimum level of performance it degrades the system performance[1]. As with the examples above, no component functions in isolation and no subsystem functions in isolation. There are always interfaces with other components, other subsystems and other systems and these have to be taken into account when designing any system.

## System requirements

When designing anything, there has to be a requirement – a definition of what the product or service has to achieve. With saleable products and services the requirements are expressed by the customer as needs, wants and expectations. Without a requirement, the product or service may not have a market and may be unsellable. However, many products create demand by offering new benefits that no one thought were needed. If the products are successful, before too long everyone wants them. Such was the case with television, radio, video recorders, cameras etc. Before the camera existed no one said: 'I wish I had a camera to record this event.' They used whatever was available, such as pencil and paper. Once the camera was available it became an essential tool for many and demands for its improvement followed. Customers were asked what they would like a camera to do and technology provided the solutions, often ahead of the actual needs.

When we seek any product or service we have five basic needs: purpose, functions, features, conditions of use and constraints.

### *Purpose*
The first need is the product's purpose – what we want the product to do. In the case of a camera, we want it to take pictures on roll film. (We do have choices, as cameras are also available for taking pictures on plate, video tape or computer disk.)

---

[1] A principle developed by Eliyahu M. Goldratt in his book *The Goal* (1984).

### Function
The second needs are the functional needs – what functions we want the product to possess, such as a viewfinder, focus adjustment, film advance, aperture adjustment etc. Although we would be unlikely to purchase a camera without any of these functions, we may be looking for these functions to be automated.

### Features
The third need is the features we desire in the product. We may want the camera for close-up work, for capturing high-speed images, for attachment to a microscope, a telescope, a tripod etc.

### Conditions of use
The fourth need is conditions of use. We may want to use the camera under water, in dusty conditions, in outer space, in a studio etc. These conditions affect how the camera functions.

### Constraints
The fifth needs are the constraints we impose. The camera needs to be safe, reliable, require no maintenance, use recyclable materials, sell within a certain price range etc. These affect both the camera design and its construction.

## Quality system requirements

Let us now apply the same principles to the quality system.

### Purpose of quality system
The purpose of the quality system is to ensure that the products and services supplied by the organization meet customer requirements and to enable the company to achieve its stated quality objectives.

### Functions of a quality system
In a specific quality system, the functions would be the core business processes.

### Features of a quality system
The minimum features of a quality system are those specified in ISO 9001 or ISO 9002 but we may require others – such as information security, contamination control, traceability to raw materials – depending on specific critical success factors.

### Conditions of use for a quality system
The conditions of use equate to the scope of the system, to what products and services it applies and where it is to be applied. However, we may impose other conditions such as: that the documentation be accessible on a computer network, that the documents be in two languages, that the system be portable to other locations, etc.

## Constraints on a quality system

The constraints on the quality system would cover such aspects as:

| | |
|---|---|
| **Robustness** | The ability to withstand variation in the way operations are carried out without system breakdown |
| **Complexity** | The number of interconnections, routings, pathways, variations, options, alternatives etc. which give rise to multiple procedures |
| **Maintainability** | The ease and economy with which system changes can be made |
| **Reliability** | The extent to which the system produces consistent and predictable results |
| **Flexibility** | The ease with which the system handles changing circumstances |
| **Vulnerability** | The extent to which the system is dependent upon certain resources |
| **Consistency** | The extent to which the documented system unifies communication both within itself and within the organization in which it is employed |
| **Compliance** | The extent to which the system complies with the requirements of ISO 9000 or other prescribed requirements |
| **Usability** | The ease and economy with which the system enables users to determine the right things to do and to do these things in the right way the first time and every time |
| **Traceability** | The ease and economy with which the system enables information to be traceable to the governing requirements and vice versa |

These needs should be documented in a Quality System Requirement Specification and this document used to direct the system design and to evaluate system effectiveness. You don't have to go to such lengths – in fact most companies don't. They simply use ISO 9000 as their set of requirements but you may need to meet other requirements to achieve your quality objectives. However, you will undoubtedly have other needs and no matter how simple or complex these are it is good practice to document them so that those designing the system have a point of reference. What it amounts to is that you design a quality system as you would design a product. You may design products without specified requirements and as a consequence have to redesign them several times. The Quality System Requirement Specification is the quality system equivalent to the design input document mentioned in clause 4.4.4 of ISO 9001. It is used both as a design input and as a basis for determining the effectiveness of the system following its installation.

## Composition of the quality system

The quality system consists of the organization, processes and resources needed to implement the quality policy and achieve the quality objectives, as covered in Chapter 2, *TQM relationships to ISO 9000*. The composition of each of these elements is illustrated in Figure 5.1.

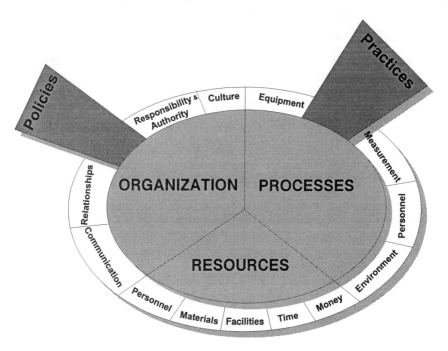

Figure 5.1 *Composition of the quality system*

The organization consists of the structure, roles and responsibilities that will cause the system to function[2]. The system doesn't function without people and it is what these people are required to do and their interrelationships that makes the system work. You can write procedures until you are blue in the face but without people being assigned to jobs that require their use, these procedures are no more than books on a shelf. People need a reason to read the books and it is the organization that does this. The inherent values in the culture shape behaviour, and behaviour will dictate what people do and where they place their priorities.

Resources include everything needed to accomplish a task and include the measures taken to secure and develop the resources to deliver the desired performance. By applying the resources and the organization to processes, inputs are converted into outputs of added value.

[2] See also ISO 8402.

## System design strategies

### The options

There are several options to choose from when designing a quality system:

- Follow the standard

- Document what you do

- Design business processes

### 'Follow the standard' method

ISO 9001 specifies clearly what documents are required to describe your quality system. However, they are only topics or subjects to be described and may not reflect the nature of your business. This is particularly true in the service industries. The standards were written for manufacturing industry and are therefore written in a language common in that industry. A list of documents is given in Annex A. Some will not be applicable because you don't do certain activities such as design, servicing, installation or receive customer supplied product. The titles of others you may wish to change to reflect your business, for example: Control of Nonconforming Product may be titled *Problem Management* and Contract Review may be titled *Service Level Agreements*. Whether you cover more than one topic in a document is a matter of choice. How will the document be referred to? If it will be used in several areas, then combining topics may require you to be specific when referencing it – users may otherwise confuse your requirements. Separating topics gives you flexibility and simplicity and avoids ambiguities. An auditor may ask the sales manager for your customer complaints procedure. The sales manager knows there is one but cannot find it because it is part of the corrective action procedure. The auditor draws the conclusion that the sales manager is unfamiliar with the system and proceeds to search for more examples.

The advantage of this approach is that the documentation produced matches the standard and it is therefore easier for the auditor to verify compliance.

The disadvantage of this approach is that you may have produced a 'bolt on' system where the documents are generally descriptive of what you do but are not used. It doesn't meet the intent of the standard – the intent being to secure customer satisfaction through the prevention of nonconformity at all stages from design through to servicing. The operating procedures or work instructions become the documents that are used to direct work. There are obviously some exceptions, such as the auditing and training procedures but, in general, these procedures and manuals reflect what the auditor wants to see and not what the business needs. Collectively the documents do not describe a sys-

## 122 System design

tem because ISO 9001 is an assurance standard and specifies only the minimum requirements for quality assurance. There was no intention that all the documents specified should fit together as a system with outputs from one procedure being the inputs to another. It for this reason that ISO 9004 is recommended in ISO 9000-1 as being the preferred standard for developing a quality system.

Another disadvantage is that the level of detail varies significantly. The quality manual becomes a copy of the standard with section headings identical to the standard and policies that reiterate the requirements of the standard. Manuals of this type vary in the depth to which they address the requirements. There are over 300 requirements in ISO 9001 and yet many manuals limit the response to the elements and sometimes the clauses but hardly ever to each individual requirement. The argument for limiting the response is simple. 'Many of the requirements are addressed in the procedures, therefore we do not need to repeat in the manual what we have written in the procedures.' For auditing purposes, this makes possession of the procedures essential to carry out the documentation audit.

### FOLLOW THE STANDARD

| Advantages | Disadvantages |
|---|---|
| • Meets the standard | • Doesn't create a system |
| • Auditors like it | • Doesn't match the business |
| • Takes little time | • Documents not used |
| • Inexpensive | • Becomes quickly obsolete |
| • Functionally independent | • Doesn't involve everyone |
| | • Focuses on quality assurance not quality management |
| | • Doesn't meet the intent of the standard |

### 'Document what you do' method

A commonly held belief about ISO 9000 is that the requirements can be summarized as follows: Document what you do, do what you document and prove it. If you document what you do now, it will not necessarily create a system. It depends on how you describe what you do:

a) As a collection of activities

b) As a series of departmental functions

c) As a series of interconnected processes

This approach causes a number of problems. Firstly the 'you' is not defined. When sending out this message to the managers, the consultant or CEO when referring to 'you' means the company. However, the managers each go away and tell their staff to 'document what you do'. The 'you' now becomes not the departments but the individuals. Sometimes the managers realise that 'you' does not mean individuals and produce departmental or function manuals.

The advantage of this approach is that it captures the tasks that are currently being carried out and generally describes how they are performed. However, the level of detail varies significantly. Some procedures are written as lists of rules and leave significant gaps in understanding as a consequence of the writer making

> **DOCUMENT WHAT YOU DO**
>
> | Advantages | Disadvantages |
> |---|---|
> | • Easy to do | • Doesn't create a system |
> | • Reflects current practices | • May be noncompliant |
> | • Everyone involved | |
> | • Doesn't tie you to a specific standard | • Doesn't account for automated processes |
> | | • Produces lots of paper |
> | | • Functionally dependent |
> | | • Takes longer to create |

assumptions. A likely response might be: 'After all, there is no need to spell everything out as we all know where to get the information and where to send the output, what the terms mean and what the sequence is. We also don't want to be constrained to a specific method as we like everyone to use their initiative.' Another consequence is that the writers put down everything regardless of there being a requirement in the standard. Procedures are written for administrative functions as well as product-related functions and the net result is a mountain of paper that by the time it is released has become history because organizational and process changes have occurred.

By documenting what you do you might be making the assumption that what you are doing is OK because it meets the standard when, in fact, this could be far from the truth. There are likely to be tasks that the standard requires that you are not doing, such as specifying the recording requirements before you perform the inspections (ISO 9001 clause 4.10.1), or determining you have the capability to meet customer requirements before you accept the order (ISO 9001 clause 4.3.2).

Documenting what you do also tends to capture only human activity. In an automated plant, humans may do very little apart from monitor the processes. The process control is automated. The description of the process control is probably included in the plant manuals but unless you reference these manuals in your system description then you have not addressed the requirements of the standard. Such descriptions will not be traditional procedures, but automated procedures that need to form part of your system. Why? Because you need to understand how your products and services are generated, how they are controlled, what causes things to happen and what prevents undesirable events from happening. Without such knowledge you cannot be managing. Your actions and decisions may appear sound but unless you know what you are doing, you are not in control. Imagine deciding to replace some machinery and, after its installation, you find that the level of customer complaints has risen. Why? Because your new machinery did not have some automatic feature that you were relying on to deliver good product. In reality, you may be fortunate and someone in your company might inform you of the mistake before you commit yourself to the installation. However, what if this change follows a period of staff changes and the key individual is not around to alert you to the dangers?

> ❑ The problem with placing all your trust in staff to control the outcomes of the business is that you never know what it is that they do that you are relying on until they no longer do it.

With the 'document what you do' method, you don't create a system but describe all of the components. Now, if that is what you have done you can still create a system from the components by using the business process methodology.

### 'Design business processes' method

#### All work is a process
All work is performed through processes. Work commences with an input and proceeds through a series of steps to produce the output using tools, equipment and so on. A process can be as small as a series of activities for accomplishing one task or as large as a factory producing raw materials.

Work tends to be performed by passing information or product from one person to another or one machine to another. In reality, work is not accomplished vertically in an organization. Authority passes downwards, responsibility upwards and work horizontally, as illustrated in Figure 2.5.

There will of course be some vertical relationships (such as the manager passing instructions to the subordinates and the subordinates passing decisions to the manager) but when the subordinates begin to work they will liaise with colleagues in other functions and will not pass information up and over the fence into the next function before getting a response.

#### Business processes and work processes
Business processes are groups of work processes that deliver business outputs. Each business process consists of a combination of work processes that are performed by individuals or teams of individuals. These teams may be functional or multifunctional, such as a project team comprising members from several functions.

#### The methodology
Unlike the other two methods, the business process method does not start with the standard or documenting what you do. It starts with a model of the business – a model

indicating the business processes and their relationships. Each business process is analysed in terms of the work processes that turn the inputs into outputs and each work process is again analysed in terms of the tasks performed on the information or product as it passes through the process. Each task is then analysed to determine the individual actions and decisions that are needed to perform the task and who performs them.

> **DESIGN BUSINESS PROCESSES**
>
> | Advantages | Disadvantages |
> |---|---|
> | • Creates a system | • Takes longer to create |
> | • Matches the business | |
> | • Doesn't tie you to any standard | • Needs a common language |
> | • Functionally independent | |
> | • Everyone involved | |
> | • Encourages team work | |

At each stage the relevant requirements of the standard are assigned to each process, task and activity and where necessary new processes, tasks and activities added to create a compliant and coherent system. Only after this stage is reached are the tasks and activities documented in the form of procedures and standards. With the business process method, processes are connected through the flowcharts and the connecting lines are channels along which information or product pass. The inputs and outputs are therefore predefined and, providing you keep to the rules, you will create a system that really does cause outputs of the required quality and prevent outputs of poor quality.

The advantage of this approach is that it creates a system that causes conformity and prevents nonconformity through a chain of related processes. It provides a systematic way of identifying documentation needs and results in a system that matches the business and complies with any standards. The system can therefore be modified to match the requirements of any number of standards and regulations that are imposed on the business. Unlike the other two methods, it does not tie you in to a particular standard and will result in a more robust system. The documentation becomes a description of the system, resulting in a documented system not a system of documentation (as with the other two methods).

This method causes staff in different functions to talk to each other and work out together how a process converts given inputs into outputs. This process is not easily accepted as departments may not be accustomed to working together. They often see only their part of the process and do not give a hoot about what happens before or after they have done their work. But everyone contributes to the business outcomes and each contribution needs to be placed in relationship to the other contributions and not treated in isolation. One disadvantage is that it may take longer to develop such a system and some auditors may find it more difficult to relate to the requirements of the standard. Although auditors may easily recognize the first method, the business process method will in fact help auditors plan their audits more effectively and provide them with far more insight into the way you conduct your business. Another disadvantage is that it needs a common language. With the other system types, departments are free to use whatever terminology they prefer as the staff in the department will using only the documentation that relates to their department. With the business process solution, there will be documentation that is common to many departments as it will describe a process that extends across several departments. Agreement on terminology is vital to avoid misunderstanding. For instance, if you use a generic term such as 'Problem Report', each department may have a different label for this and not realize that the procedure requires an action on their part. In small organizations, changing to common terms may be less painful than in large organizations but the change has to be accomplished if the system is to be effective. The alternative is to surrender to the departmental approach where, apart from some high-level procedures, volumes of departmental procedures are created. This is more difficult to control and could well result in a fragmented system.

## System development process

### The system hierarchy

There are six levels in the hierarchy of the Quality System Design. The first level is the *system* which comprises *subsystems*. A subsystem may consist of two or more *business processes*. Where there is no subsystem, the business process is the next level down from the system. Business processes comprise *work processes* and work processes comprise *tasks* which comprise *activities*. This final division forms the basis of procedure models. The hierarchy is depicted in Figure 5.2.

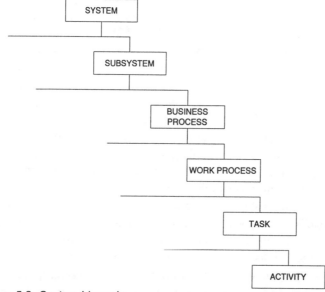

Figure 5.2 *System hierarchy*

The definitions are as follows.

## System
The *system* is the Quality System. It is the whole. It consists of all the policies, practices, resources, processes and structures that serve the achievement of quality. The system is represented by a context diagram and system model, which show the relationship between the subsystems or business processes and the interfaces to the external environment.

## Subsystem
A *subsystem* comprises a group of business processes that serve a common purpose in the organization, such as resource management, product development and quality system management. With resource management there are several types of resources, each of which may be managed through different work processes. Likewise with product development, different types of product (such as turnkey systems, software and hardware) may be managed through different work processes. The subsystem group is only necessary for modelling the system when intermediate layers are needed before identifying work processes.

## Business process
A *business process* is a special or major activity that is derived directly from the organization's purpose and mission statement and its corporate strategy. Business processes are what the organization does and include what the business needs to survive as well as to grow (for example Marketing, Design, Procurement, Servicing, Accounting, Manufacturing, Service Delivery, Maintenance). Each one of these makes a unique contribution to the business. The processes are related to the business objective that the work serves, not what job title people have or in what department they work. For example, purchasing is a business process and many people may be involved from the person who identifies the need to the person who checks the goods into the company. The business objective served by purchasing is to acquire goods at minimum cost that meet company requirements and are available when needed. The person checking the goods may perceive the job as stopping defective goods from entering the plant; the person identifying the need perceives this work as planning. They both may work in different departments but they both serve the same business objective.

A business process may be carried out by people in several different departments or sections. Business processes consist of a series of work processes combined together in such a way that they fulfil a unique purpose in the organization.

## Work process
A *work process* is a sequence of tasks which combines the use of people, machines, tools, methods, environment, instrumentation and materials to convert given inputs into outputs of added value. The output may be a product or a service. It may be information or a decision. Work processes are identifiable elements which, when combined together,

create a business process. Work processes consist of a series of 'tasks' combined together in such a way that they fulfil a certain objective. A business process may consist of a single work process and in such cases the two are the same but often a business process consists of several work processes. Some work processes may be carried out by one function but often they are multifunctional.

### Task
A *task* is a group of activities that accomplish a minor objective. Tasks are jobs to be done in carrying out a work process. Often the description given to a task is the same as that given to the related process. For example, Repair and Modify are both tasks but each is only one step in the repair process or modification process. In a work process there may be several tasks, each carried out by different people, some of whom may belong to different departments. There may also be situations where a task is executed through several functions.

### Activity
An *activity* is the smallest parcel of work. Activities are things to be done to accomplish a task. Activities consist of a series of movements or routines, the performance of which may be defined by criteria or actions. For example, one task may be to prepare a certain document and the activities may consist of collecting the input data, preparing a draft, typing the draft, issuing the draft for comment, reviewing the draft etc. up to the point where an approved document is released. Activities are carried out by individuals or teams of individuals doing the same task. In some cases, all the activities to accomplish a task may be performed by one individual and in others each activity is performed by a different person.

### Development methodology
With this *development methodology* you don't start with the standard and commence documenting your practices – in fact that is the last stage of system design. The sequence is illustrated in Figure 5.3.

This process can take several months. The first process models can be accomplished in one day but the charting and document identification tasks can take 2-3 months, depending on the availability of personnel and the clarity with which they perceive the way business is carried out

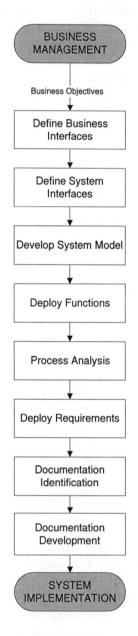

Figure 5.3
*System development process*

in the organization. Once the documentation is identified, its development can also take 2-3 months. Even with small organizations, the process can take a long time – the smaller the organization, the fewer the resources at its disposal. The larger the organization the more likelihood of resources being available but there may be more documentation to produce.

## Flowcharting conventions

There are several international standards which define flowchart conventions, many of which apply to software modelling. You can depict documents, data, storage media and other symbols but you need to bear in mind the readers and the level of understanding they have. Use symbols that are unfamiliar and they will not communicate the intended message. Limit yourselves to strict connection rules and the diagram becomes too large and complex. With these diagrams you are not designing a software program or laying out the drawings of a power station. The simplest conventions are given in ISO 9004-4 and these have been adopted in the diagrams presented in this book, with some minor modification (see Figure 5.4).

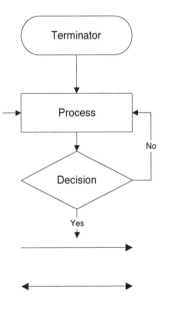

This shape represents the source of an input or the destination of an output. The source can be customers, suppliers, processes or procedures

This shape represents a process, task or activity. Input and output connections can be made on any side, but the normal flow should be top to bottom.

This shape represents a decision and the possible results are depicted on the output lines.

Lines with single arrows show the direction of flow from one shape to another.

Line with double arrows indicate two way flow between the same shapes

Text on a line identifies the nature of the product or information that is flowing between the connected shapes

Figure 5.4 *Flowcharting conventions*

## 130 System design

Beware of some circular routes when drawing your flowcharts (as illustrated in Figure 5.5).

Figure 5.5 *The circle line*

## Business interfaces

No business operates in isolation. There are always interfaces with the outside world. Through such interfaces pass information or product either into the organization or out of the organization. Many organizations are part of much larger organizations and have a relationship with other divisions and corporate headquarters as well as the end user of the product or service. Some organizations do not interface with the end user, their customer being another part of the same corporation. These arrangements need to be defined so that it is clear where the inputs come from and go to and who the customers and suppliers really are whose requirements have to be satisfied.

A model of these relationships should be produced. We will call this the Context Diagram as it places the organization in context with its surroundings. A generic context diagram is given in Figure 5.6.

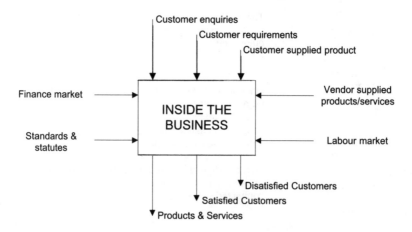

Figure 5.6 *Generic context diagram*

System design  131

In developing the context diagram you should answer the following questions for the business as a whole:

- Who are our customers and what do they supply to us? (Information, products, requirements.)

- Who are our suppliers and what do they supply to us? (Information, products, services.)

- Who are our regulators and what do they regulate? What services do we rely on from other parts of the enterprise that are outside our business?

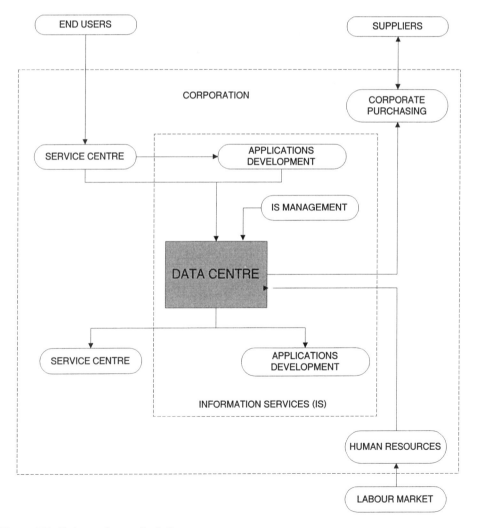

Figure 5.7  *Data centre context diagram*

## System interfaces

Within the business we may not want a quality system to encompass everything inside the business. Some parts of the business may exist for social reasons, for public service reasons; not every function in the business will serve the purchaser of your products and services. Not every stakeholder is a purchaser or user of your services. When considering what should be inside the quality system and what should remain outside, you need to establish whether the processes outside contribute to those inside. If there is no connection then ignore them. If there is a connection then they should be inside the quality system. If you use products, tools, equipment etc. from processes outside the system, the system has to provide for such inputs to be approved before being used. Processes that supply these inputs are external to the system. However, if you use products, tools, equipment etc. to develop, implement and maintain the system, then the processes for supplying these inputs should be inside the system and hence form part of the system.

An example of a data centre that serves customers within a corporation but has direct contact with end users is shown in Figure 5.7 above. In this case, the data centre chose to develop its own quality system and hence has multiple interfaces.

A further example of a context diagram is given in Figure 5.8. This is for an independent manufacturing company supplying prefabricated buildings to a housing association. The external organizations are shown as terminator boxes. These should be carried through to the lower level charts where there is a direct interface between the organization and the external body.

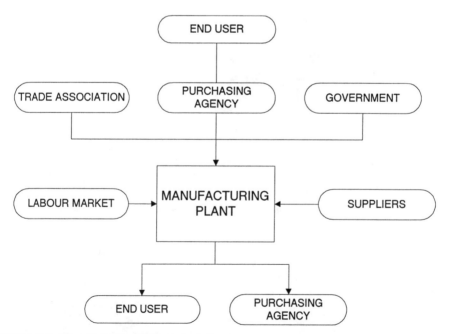

Figure 5.8 *Manufacturing plant context diagram*

## The system model

### Purpose

The *system model* is a diagram depicting the chain of business processes that convert business inputs into business outputs. Its purpose is to show the relationship between the key processes and which of these interface directly with the organizations external to the business. The system model is the general arrangement diagram of the business – the equivalent of the top-level drawing of a product. The lines connecting the processes are channels along which information or product pass.

### Creating the system model

In creating the system model, the first task is to define the first event following the receipt of a customer enquiry. This should answer the question: 'What is the first thing you do on receipt of a customer enquiry?' There may be a step before this, such as a process of attracting customers so that they make an enquiry. At the centre of the model are the core processes, those that create the products and services upon which the business depends for its survival. The next step is to define the second process in this chain and so on until the last process supplies the outcome to the customer. In the drafting stage, many steps may be identified – in fact it is good practice to reveal the intermediate steps. When you think you have defined the chain of events, you can start to rationalize the model. Some events may serve a common objective and be given a generic name (such as marketing, design, production etc.) and these can be merged to form a business process, as shown in Figure 5.9, which uses a marketing example. It is useful, at this level, to indicate the nature of the product or information that passes into the system.

### Rules of construction

The boxes in the diagram that represent processes should not carry the names of departments or functions if at all possible. Placing function names on the diagram is a distraction from the purpose of the model. People will look at it, see their department name and limit their perspective to the work of their department. This is not what you are depicting by the model. Let us take 'Marketing' as an example. There may well be a marketing department in your company but does it operate in complete isolation from everyone else? No! There are activities carried out by other departments that serve the purpose of marketing. The model does not show how departments are linked together. It shows how processes are linked together and in each business process, many departments will be involved. A common perception when drawing this model is to identify Production as a business process then place a process that is called Quality Control either as a box following Production or supporting Production. In Figure 5.10, the boxes rep-

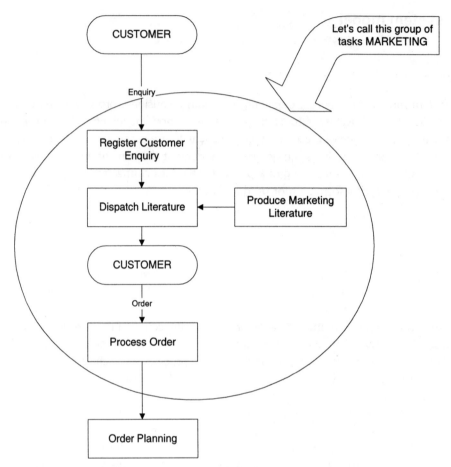

Figure 5.9  *Grouping tasks to form processes*

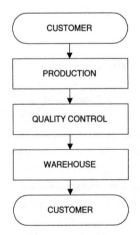

Figure 5.10  *A functional relationship*

System design 135

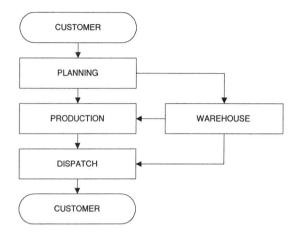

Figure 5.11  *A process relationship*

resent functions or departments. The Production Department carries out planning and is responsible for the warehouse but not quality control. In Figure 5.11, the boxes represent processes where quality control is part of the production process and warehouse and planning two separate processes. In the functional model, there is no indication that products are not made to order. The diagram seems to indicate that all orders have to go through production, whereas in the process model it shows that existing product can be taken straight from the warehouse to dispatch without having to be processed through production first. A simple change from a functional to process-based chart reveals information not previously apparent.

Changing one's mind set from function to process is one of the first hurdles to overcome. It may take some time before everyone has received and understood the message. The only way to be sure that they have understood is to get them to work in multifunctional teams and produce work process flowcharts. If these turn out like departmental charts then continue with the education until they produce charts that show how all of the departments relate to one another through the tasks they perform.

The interfaces should be represented by a terminator box indicating they are outside the system being designed. All lines should have arrows indicating the primary direction of flow. If the channel takes two-way flow, arrows at each end of the line can indicate this. Do not add every conceivable interconnection at this stage. What you should indicate are the primary channels within which may be many sub-channels. (The fact that the channels may have one lane or twelve lanes, lay-bys and refuelling stations is immaterial in this diagram!)

**136** System design

## The generic system model

A generic system model is illustrated in Figure 5.12.

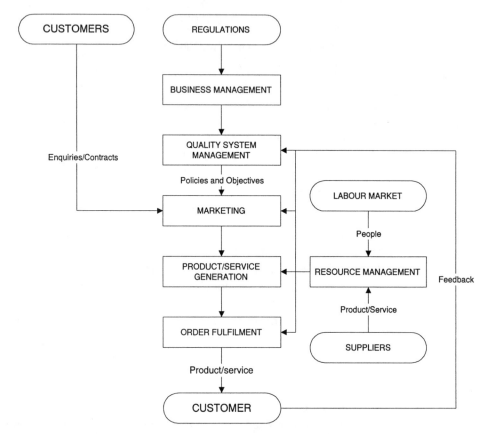

Figure 5.12 *Generic system model*

The generic system model shows the external interfaces, the inputs and outputs from these external interfaces and the business processes that convert the inputs into business outputs. This model is very simplistic but at this level it represents any business. As a working model it has no practical value; we have to expand some of the boxes to get at the nature of the business processes. Each business is different but many fall into one of several classes, depending on where the customer order enters and which processes convert orders into products and services. A series of models representing various types of business are displayed in Figures 5.13 to 5.24. For simplicity, some of the interfaces of these models have been omitted. All except the In-patient model have a marketing process, although in some health care operations there is marketing, depending on whether it is a private or public system.

You will notice that in some cases, even though only the typical ones have been included, the range of businesses that fit the model is quite large. For a specific business the model may be different but will follow the same pattern. In the catering industry, for example, there is such a wide range of catering businesses that they cannot all be represented by a single model. You may, for instance, consider that there is 'design' in a catering business because the recipes have to be created and the results tested before production commences. In another case, an advertising business may, in fact, produce the advertising material or subcontract its production so that other processes would need to be added. These models are intended only as a hypothetical guide and not as authentic models of these businesses.

When creating the system model be aware that it is an iterative process. You will not get it right at the first attempt. As the design unfolds and you gain more knowledge about the business, modifications will need to be made to the system model to make it representative of the business. The model often starts out either too simple or too complex and moves towards the optimum as the design progresses.

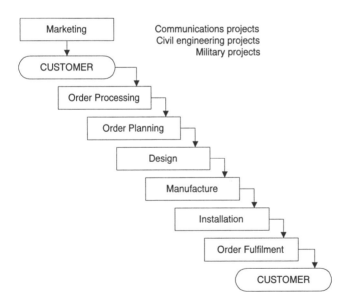

Figure 5.13 'Design, make & install to order' model

**138** *System design*

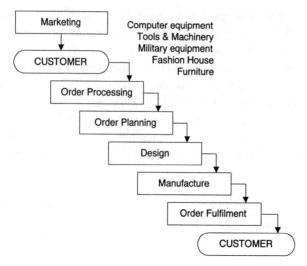

Figure 5.14 *'Design & make to order' model*

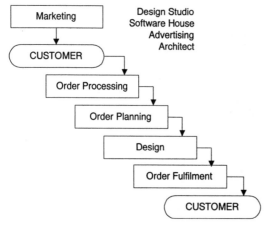

Figure 5.15 *'Design to order' model*

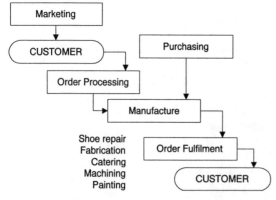

Figure 5.16 *'Make or process to order' model*

*System design* **139**

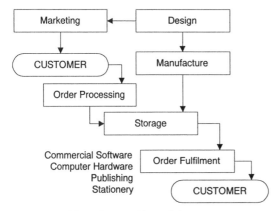

Figure 5.17 *'Make for stock' model*

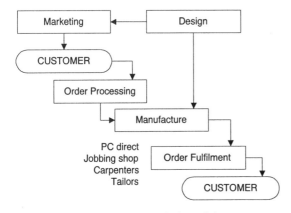

Figure 5.18 *'Manufacture to order' model*

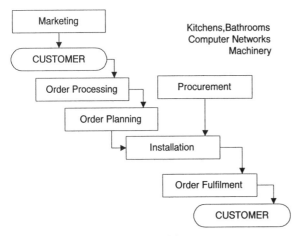

Figure 5.19 *'Install to order' model*

# 140 System design

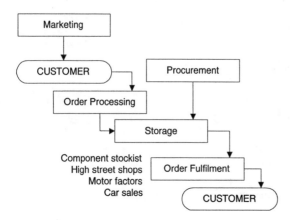

Figure 5.20 *'Wholesale & retail business' model*

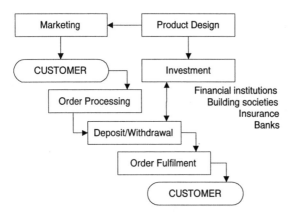

Figure 5.21 *'Financial business' model*

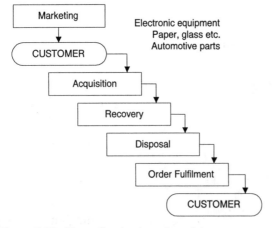

Figure 5.22 *'Recycling business' model*

*System design* 141

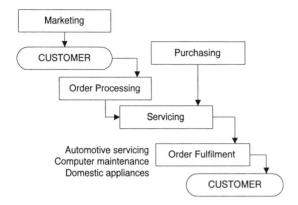

Figure 5.23 *'Servicing business' model*

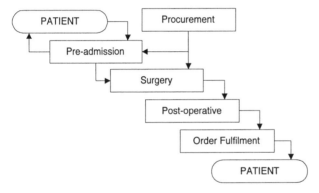

Figure 5.24 *In-patient surgery*

## Function deployment

In creating the system model the functions or departments that are involved should have been omitted because only the processes that are carried out have been identified. To ensure that you have not overlooked any key functions a matrix of functions against the business processes can now be developed. This matrix will show the contribution each function provides to each process. When placing the bullets in the boxes, each bullet should represent an action or decision taken by the function, not mere presence at a meeting or receipt of information. When you chart the work processes there should be a box in the process flow with a label identifying the function carrying out the task. Some functions may contribute to only one process whereas others may contribute to many processes (an example is given in Table 5-1). If this stage is too coarse to make the matrix meaningful then an alternative is to use the same technique when the process charts are created.

| FUNCTION | PROCESSES | | | | | |
|---|---|---|---|---|---|---|
| | Business Management | Quality System Management | Marketing | Resource Management | Product/Service Generation | Order Fulfilment |
| Executive Management | ● | ● | ● | ● | | |
| Marketing | | ● | ● | ● | | |
| Sales | | ● | ● | ● | | ● |
| Systems Engineering | | ● | ● | ● | ● | |
| Product Development | | ● | ● | ● | ● | |
| Environmental Test | | ● | | ● | ● | |
| Drawing Office | | ● | | ● | ● | |
| Computer Centre | | ● | | ● | ● | |
| Software Development | | ● | | ● | ● | |
| Development Manufacture | | ● | | ● | ● | |
| Production | | ● | | ● | ● | |
| Purchasing | | ● | | ● | | |
| Quality Control | | ● | | ● | ● | ● |
| Quality Assurance | ● | ● | ● | ● | ● | ● |
| Finance | | ● | ● | ● | | ● |
| Dispatch | | ● | | ● | | ● |
| Maintenance | | ● | | ● | | |
| Service | | ● | | ● | ● | ● |

Table 5-1 *Function Deployment Matrix*

## Analysing business processes

### The generic process

All processes have certain generic features. There are always inputs and outputs and, following inputs, some preparation is needed before work proceeds. During and after the work, checks are carried out to verify the work is being performed correctly and, should errors be detected, action is taken to correct them. When the work is deemed acceptable it is released. This is the process of control and fits all work situations. It also fits any control situation whether the work is performed at the enterprise level, the business level or the operational level. At the enterprise and business level the controls are effected through information whereas at the operational level the controls are effected through a number of techniques, as indicated in the Control Pyramid[3] (see Figure 5.25).

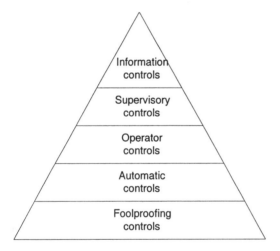

Figure 5.25 *The Control Pyramid*

If the process managers are smart they will also take action to prevent any problems recurring but they may do this periodically rather than after each job. This last step is the key to process control (as opposed to work control). Work control controls the output. Process control controls the process that produces the output and will therefore ensure all outputs are correct. A diagrammatic representation of this generic process is given in Figure 5.26.

This diagram has some common features with the Deming Cycle. His PDCA or 'Plan, Do, Check, Act' cycle is laid out as a flowchart (not a circle). The 'Act' part of it is divided into two – action to fix the problem and action to prevent its recurrence. Both are

---
[3] *Managerial Breakthrough*, J. M. Juran (Second Edition, 1995).

## 144 System design

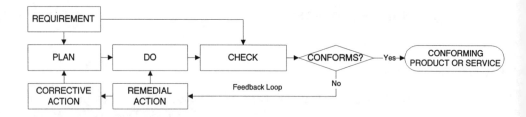

Figure 5.26 *Generic process*

essential to maintain control. Juran defines the sequence rather differently[4]. Juran states that control consists of the following:

1 Choosing the control subject (part of the Requirement)

2 Choosing a unit of measure (part of the Requirement)

3 Setting a standard value (part of the Requirement)

4 Creating the sensing device (part of the Planning)

5 Mobilizing for measurement (part of the Planning)

6 Conducting actual measurement (part of the Check)

7 Interpreting the difference (part of the Check)

8 Decision making and acting on the difference (part of the Remedial Action)

Juran omits the 'doing' stage as what he defines is 'control' rather than a generic process. Deming uses the PDCA as a means of improvement: planning change and then following a continuous cycle of year-on-year improvements. Nevertheless, both concepts are valid for any type of work and match each other rather well. The only omission appears to be the 'corrective action' stage. If this stage were not to be taken, then the output of the process would not change. It is only when the solution eliminates the cause of the problem that the performance of the process returns to normal. The remedial action deals with the product of the process and the corrective action the performance of the process. In both Deming's and Juran's models, corrective action is the action to which they refer. Process control therefore has to include a 'corrective action' stage otherwise the process is not under control.

---

[4] *Managerial Breakthrough*, J. M. Juran (Second Edition, 1995).

We need to understand the generic process to determine whether processes are under proper control. This is yet another reason why the 'Document what you do' method is not always helpful.

## Process analysis methodology

Following creation of the system model, each of the business processes can then be analysed and the work processes defined down to a level where documentation needs can be identified. At each stage, the requirements of the standard are deployed so that those identified in the system model deployment matrix are linked to tasks in the work process charts and deployed further to individual activities. In this way, the charts demonstrate traceability of requirements forwards from the standard to the point of implementation and backwards to the governing requirements. Requirement traceability modelled in such a way facilitates the change control process. When processes need to be changed, the charts and matrices will indicate both the governing requirements and the interfaces that may be affected by the change. When working at the level of procedures, it is easy to overlook governing requirements and interfaces and make changes that create conformity and integrity problems. With only one governing standard, the chances of error are less, but when the system needs to comply with many governing standards, such errors are easily made and a route map becomes a necessity.

## Process description

Before charting the work processes that comprise a business process, the description of the business processes should be defined and agreed. The process descriptions enable you to determine which tasks are included and which are excluded. If you work blindly, then there is a chance that you will include tasks that are really part of other processes. The descriptions are also an effective means of unifying understanding and agreeing a common language.

The process description has two components: the purpose of the process and the boundary conditions. A typical process description would be of the form:

> The purpose of ... process is to ... It commences with the receipt of ... from ... and ends with the release of ... to commence the ... process.

## Charting processes

Each work process consists of tasks and sometimes other work processes. Using the generic model as your guide, chart the sequence of events that follows receipt of the input from the previous business process or (if it is the first process) from the customer.

**146** *System design*

All work requires an input to commence. In many cases the input is either a product or a piece of information. Even with time-dependent tasks, the task commences with the release of the time schedule that stipulates when the task is to be performed. Ask the question: 'From where do you get your instructions to trigger this process?' The decomposition of processes is illustrated in Figure 5.27. Here we take one of the business processes of the System Model, model the work processes that comprise this business process and then expand each task in the work process into activities to form a further flowchart. The inputs and outputs should be shown at each stage in the decomposition. This illustration is symbolic; the tasks may interface with other processes and there may be decision tasks as well as action tasks. The tasks in the business process chart are likely to cross departmental boundaries so that the responsibility for each task is different. In the work process chart, the responsibility for each activity may be the same but could be different.

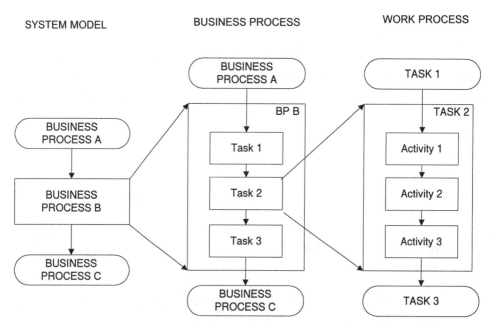

Figure 5.27 *Process decomposition*

## Charting methods

### Process charts
The process chart depicts the steps in the process from input to output by shapes following one after the other either vertically or horizontally and identifies responsibilities for each step. An example of the business process chart for purchasing products is shown

System design 147

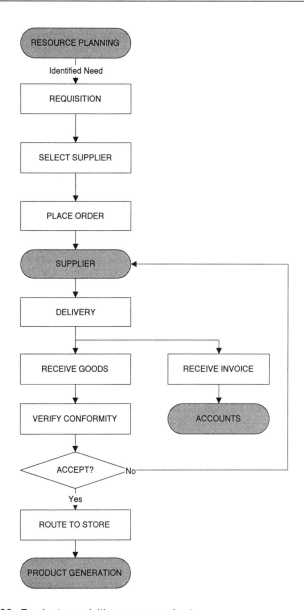

Figure 5.28 *Product acquisition process chart*

in Figure 5.28. Product purchasing is a type of Resource Acquisition and constitutes a business process. Although Resource Management is a subsystem comprising several business processes (see Figure 5.30), Resource Acquisition is a business process itself, as there is more than one type of resource to be acquired by a company and each through a different process. In Figure 5.28 the work processes are identified by the boxes and may result in further flowcharts. This chart represents the first iteration of the process showing the basic tasks. It does not show compliance with ISO 9001 at this stage. Further tasks need to be added and this is addressed later in this chapter. In some organizations

**148** *System design*

this may be the end of the decomposition before documentation is identified. In others there may be several purchasing processes and therefore an intermediate level is needed. This level we call a subsystem.

The advantage with this approach is its simplicity. However, there is little information on the chart other than tasks, sequences and interfaces. There is no description of how the tasks are performed, the identity of the documentation used or produced for each task. This can be added within or alongside the boxes but it may enlarge the chart beyond a single page. Another weakness is that if you add the responsibilities they would need to align with the boxes making it difficult to assign responsibilities for parallel routes. For example, in Figure 5.29 two tasks are shown as parallel tasks. Should the function or role responsible for each be different, either you need to place the name of the function either side of the chart or move down one of the boxes to show responsibilities along one side. This problem does not arise with the cross-functional chart.

### Cross-functional chart
The cross-functional chart depicts functions vertically and depicts the steps in the process from input to output according to the responsibilities horizontally. An example of this type of chart is given in Figure 5.29.

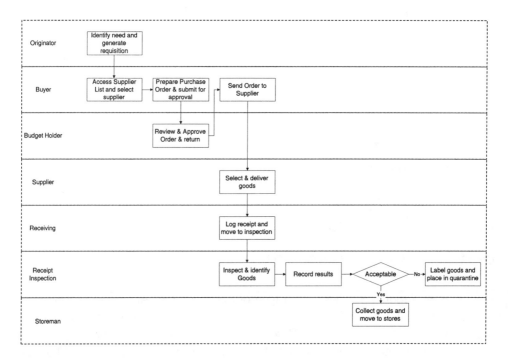

Figure 5.29 *Cross-functional chart*

The advantage of this approach is the clarity with which the responsibilities are defined and how the interfaces between functions are displayed. The approach is suitable for process analysis to identify bottlenecks and seeking ways in which the number of interfaces can be reduced. However, the amount of detail that can be shown on one sheet of paper is limited as the flow is horizontal. If one function performs many sequential tasks, it can result in lots of blank space on a page. If one function only performs one task a complete line is used, thereby reducing the space available to depict other functions. The chart can be drawn vertically to gain more space but this may limit the number of functions on each page. Naturally, if the diagrams are drawn using a flowcharting tool, their length is immaterial but consideration should be given to the user! It is extremely difficult to navigate through a diagram of several pages unless a hierarchical or layering technique is employed that enables users to quickly get to the point at which detail is needed. A software tool that does this very well is TeamFlow®. Functions can be moved by the 'drag and drop' feature. Layering of charts can be accomplished and documents can be linked, making this a very capable process-charting tool. One big disadvantage with cross-functional charts is chart maintenance. Change what an organization does and you may have to redraw the whole chart. Whereas with the process chart you merely change the names alongside the affected boxes. With TeamFlow®, changes of this nature are accomplished very easily but, nevertheless, it is a manual process. The flowchart will not automatically be redrawn when you change the organization chart.

As with the process flowchart, adding document references and other data will clutter the chart.

## Subsystems

Subsystems are business processes that comprise a number of high-level work processes. Resource Management is one example and is illustrated in Figure 5.30. Another example is a production subsystem for a product recycling plant. This comprises several other major processes, as illustrated in Figure 5.31.

In Figure 5.31 each of the boxes is a process that may consist of several work processes. The connections indicate the flow of product between processes. The acquisition process acquires unwanted product from customers which passes through a distribution process where segregation takes place. Following distribution to several distinct processes, product and materials are recovered for subsequent disposal. The subsystem concept is a convenient way of simplifying what might become very complex flowcharts should we attempt to chart all the processes in one diagram.

**150** *System design*

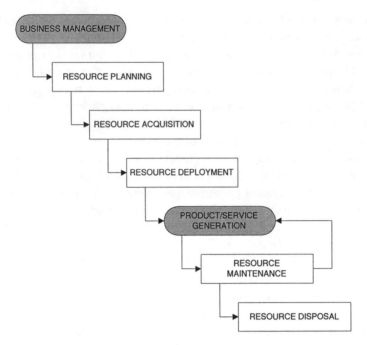

Figure 5.30 *Resource Management subsystem*

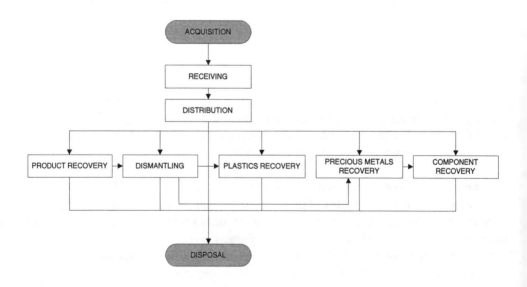

Figure 5.31 *Production subsystem in a recycling plant*

## Task analysis

Charting process flow is an iterative process. The best way of producing the charts is with a group of people in a brainstorming session. In this way the first attempts will soon be changed as more information is revealed. One way of revealing information is to carry out a task analysis (see Figure 5.32). The analysis addresses each box on the process chart and poses several questions. The results lead to changes in the flow, new inputs, outputs or even new tasks!

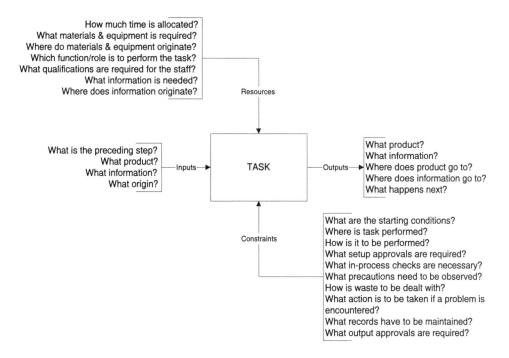

Figure 5.32 *Task analysis*

## Deploying the requirements

After creating the system model, associated business process charts and work process charts, the regulations that apply to each of the processes can be deployed in a matrix similar to the Function Matrix. This matrix can be as detailed as you like. At the higher level, only the elements of the standard are included but as an element may cover several quite different requirements, it may be more useful to include all the clauses as well. Although you can combine all regulations into one matrix (especially if you use a database or spreadsheet), for presentation purposes one matrix for each standard is preferable. Alternatively, you can display the clause numbers on the diagrams themselves. This is useful in instructional sessions as it is graphical but the diagram can get a little overcrowded if more than one standard is being addressed. Where there are many standards relevant to your business a matrix listing all the standards against the processes would be a useful guide to the system. An example of a Requirement Deployment Matrix is given in Table 5-2.

| CLAUSE | TOPIC | Business Management | Quality System Management | Marketing | Resource Management | Product/Service Generation | Order Fulfilment |
|---|---|---|---|---|---|---|---|
| 4.1.1 | Quality policy | ● | | | | | |
| 4.1.2.1 | Responsibility and authority | ● | ● | ● | ● | ● | ● |
| 4.1.2.2 | Resources | ● | | | ● | | |
| 4.1.2.3 | Management representative | | ● | | | | |
| 4.1.3 | Management review | | ● | | | | |
| 4.2.1 | Quality system | | ● | | | | |
| 4.2.2 | Quality system procedures | | ● | | | | |
| 4.2.3 | Quality planning | | | ● | ● | | |
| 4.3.1 | Contract review procedures | | | ● | | | |
| 4.3.2 | Contract review | | | ● | | | |
| 4.3.3 | Amendment to contract | | | ● | | | |
| 4.3.4 | Contract review records | | | ● | | | |
| 4.4.1 | Design control procedures | | ● | | | | |
| 4.4.2 | Design and development planning | | | | ● | | |
| 4.4.3 | Organization and technical interfaces | | | | | ● | |
| 4.4.4 | Design input | | | | | ● | |
| 4.4.5 | Design review | | | | | ● | |
| 4.4.6 | Design output | | | | | ● | |
| 4.4.7 | Design verification | | | | | ● | |
| 4.4.8 | Design validation | | | | | ● | |
| 4.4.9 | Design changes | | | | | ● | |
| 4.5.1 | Document and data control procedures | ● | ● | ● | ● | ● | ● |
| 4.5.2 | Document approval and issue | ● | ● | ● | ● | ● | ● |
| 4.5.3 | Document changes | ● | ● | ● | ● | ● | ● |
| 4.6.1 | Purchasing procedures | | | ● | ● | | |
| 4.6.2 | Evaluation of subcontractors | | | | ● | | |
| 4.6.3 | Purchasing data | | | | ● | | |

Table 5-2 *ISO 9001 Requirement Deployment Matrix*

System design  153

| CLAUSE | TOPIC | Business Management | Quality System Management | Marketing | Resource Management | Product/Service Generation | Order Fulfillment |
|---|---|---|---|---|---|---|---|
| 4.6.4.1 | Supplier verification of subcontractors | | | | ● | | |
| 4.6.4.2 | Customer verification of subcontracted product | | | | ● | | |
| 4.7 | Control of customer supplied product | | | | ● | ● | ● |
| 4.8 | Product idenitification and traceability | | | | ● | ● | ● |
| 4.9 | Process control | | | | | ● | ● |
| 4.10.1 | Inspection and testing procedures & records | | | | ● | ● | ● |
| 4.10.2 | Receiving inspection and testing | | | | ● | | |
| 4.10.3 | In-process inspection and testing | | | | | ● | |
| 4.10.4 | Final inspection and testing | | | | | ● | ● |
| 4.10.5 | Inspection and test records | | | | | ● | ● |
| 4.11.1 | Control of inspection, measuring and test equipment | | | | ● | ● | ● |
| 4.11.2 | Control procedure | | | | ● | ● | ● |
| 4.12 | Inspection and test status | | | | ● | ● | ● |
| 4.13.1 | Nonconforming product control procedures | | | | ● | ● | ● |
| 4.13.2 | Nonconforming product review and disposition | | | | ● | ● | ● |
| 4.14.1 | Corrective and preventive action | ● | ● | ● | ● | ● | ● |
| 4.14.2 | Corrective action | ● | ● | ● | ● | ● | ● |
| 4.14.3 | Preventive action | | | ● | ● | ● | |
| 4.15.1 | Handling, storage, packaging, preservation and delivery | | | | ● | ● | ● |
| 4.15.2 | Handling | | | | ● | ● | ● |
| 4.15.3 | Storage | | | | ● | ● | ● |
| 4.15.4 | Packaging | | | | ● | ● | ● |
| 4.15.5 | Preservation | | | | ● | ● | ● |
| 4.15.6 | Delivery | | | | | | ● |
| 4.16 | Control of quality records | ● | ● | ● | ● | ● | ● |
| 4.17 | Internal quality audits | | ● | | ● | | |
| 4.18 | Training | ● | ● | ● | ● | ● | ● |
| 4.19 | Servicing | | | | ● | ● | |
| 4.20.1 | Identifying need for statistical techniques | | | | ● | | |
| 4.20.2 | Statistical techniques application procedures | | | | ● | ● | ● |

Table 5-2  *ISO 9001 Requirement Deployment Matrix (continued)*

Your matrix may look quite different as some clauses might not apply to your business and the business processes will of course be those of your organization, not those taken from the generic model. This matrix is a useful way of indicating the applicability of the requirements. Many quality manuals have virtually blank pages with headings taken from ISO 9000 and the words 'This section is not applicable'. This seems to be a waste of paper and in fact indicates that the manual is really not a description of your quality system but a response to the standard. By placing this matrix at the back of the manual you fulfil both needs.

## Process modification

### Adding the controls

The charts produced should reflect how the current business and work processes are connected and what the interfaces are. After analysing the Requirement Deployment Matrices, you should be able to determine whether you need to add any additional tasks. The problem with the requirements is that some are basic, like control of nonconforming product, and others are detailed, like the requirements for purchasing data. The basic requirements should be shown on a chart, whereas the detail will not appear on a chart but on a form or in a standard.

Consult the generic process model to establish also whether there are planning tasks, checking tasks and feedback loops for remedial action. Some of these may not be required by the standard but nonetheless are necessary to maintain control over the processes.

Now examine the chart for where data needs to be collected for process-monitoring purposes (for example, time through the process, numbers of errors etc.) so that corrective action can be taken. In the process chart of Figure 5.33 (which is a revised version of the chart in Figure 5.28) data is taken from receiving inspection for analysis. The results are then used to update the register of suppliers so that subsequent users of the list will select only acceptable suppliers or be alerted to poor suppliers. The feedback loop thereby provides for corrective action.

Along each connecting line flows product or information. The identity of the specific product or carrier of information should be shown on the chart. If approval is required before work commences then the decision diamonds should be added. Similarly if approval is required before the output is released these decision stages should be shown. The interfacing processes should be shown as rounded rectangles with the name of the process not the name of the task within the process.

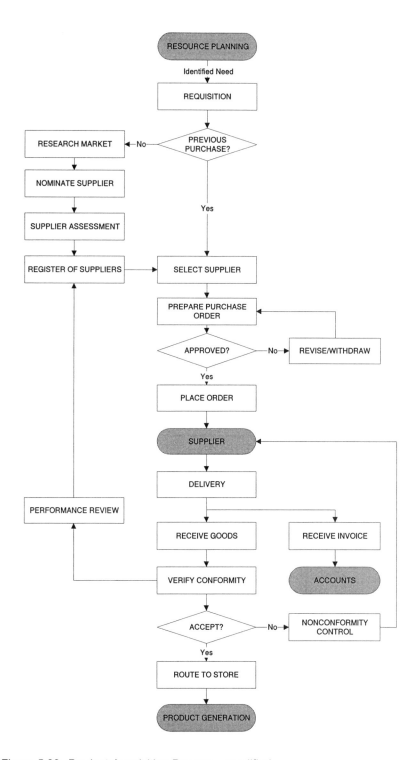

Figure 5.33 *Product Acquisition Process – modified*

**156** System design

An example showing the effect of deploying the requirements of ISO 9001 to the Resource Acquisition (Products) Process is given in Figure 5.33.

In Figure 5.33, the following additions have been made:

1. Suppliers are now selected on the basis of meeting requirements (ISO 9001 clause 4.6.2a).

2. New suppliers are now subject to assessment (ISO 9001 clause 4.6.2b).

3. Records of acceptable subcontractors are now in place as a basis for supplier selection (ISO 9001 clause 4.6.2c).

4. Purchasing documents are now approved before release (ISO 9001 clause 4.6.3).

5. Supplier data is now analysed to detect actual and potential problems (ISO 9001 clauses 4.14.2 and 4.14.3).

6. Nonconformity controls are now in place (ISO 9001 clause 4.13).

Other requirements such as the records requirements of 4.16, the purchasing data requirements of 4.6.2, etc. will be addressed in the operating procedures.

### Separating the present from the future

When charting processes there is a temptation to draw what should be rather than what 'is'. Resist this temptation! It is difficult enough to gain agreement on what is without aiming for agreement on what should be. When the system is installed, you can commence gradual or step improvements by examining better ways of doing things. If you are starting on a greenfield site, obviously you can design a system that depicts the way you want the system to be, as there is no existing system to work with. You should make it clear at the outset to all involved which approach you are taking, otherwise you will be constantly troubled by differences in understanding and impacted by delays.

### Resolving concerns

When deploying the requirements or applying the generic process model, you may come across a situation where the current processes require change to be compliant. There may also be instances where information has no carrier and is passed verbally. Assess whether such information is vital for process predictability. Does the process output depend on this information? If so, it should be documented in some form. Remember that 'the problem with placing all your trust in staff to control the outcomes of the business is that you never know what it is they do that you are relying on until they no longer do it'.

You may find you need to define responsibilities or authority where none were defined previously, or to set up a new organization to carry out a process you have not performed before. Typical examples are the internal auditing and management review requirements of ISO 9001. These issues need to be promptly resolved and one method of tracking such issues is to use a Current Concerns Report. This report can be used to gather data at the commencement of the project in order to capture the status quo, the baseline from where you are going to advance. As issues arise they are added to the report and the action item log is used to track their resolution. The reason for using two reports rather than combining them into one report is that no single action may be possible for some concerns. They may be cultural issues. However, they need to be recorded. By testing opinion later you may see a distinct change in attitude and be able to remove the concern. Not every concern requires action. Some will decline with time, others will require immediate action, others will require teamwork. A concern is something that is preventing success.

Another solution is to set up a problem management database for capturing any problems through the project and on into implementation. This problem management database can be used to implement remedial, corrective and preventive action procedures (see Chapter 4).

## Using software tools

System design is a diagrammatic process, creating flowcharts and matrices. There are many flowcharting packages available for such tasks and each varies in the features offered. Here are some options and the advantages and disadvantages.

### Drawing tools in word processors

Many of the larger word-processing packages incorporate drawing tools but they are fairly basic. They allow you to draw boxes but not diamonds other than freehand and the lines don't snap to the anchor points, so when you move a shape you have to reposition the lines manually. However, grouping features allow you to combine shapes before you move them. Formatting is basic with a limited range of styles, arrows and shapes. Such tools were not designed for flowcharting and hence are labour intensive.

## Flowcharting tools

There are many flowcharting tools available each with advantages and disadvantages. Tools such as Visio® are graphically based, whereas allCLEAR® is text based. Both are very flexible and not focused on any particular style of chart. TeamFlow® is a graphical tool that is very easy to learn and combines the use of organization charts, processes and Gantt charts to produce cross-functional flowcharts. The graphical tools provide drag and drop features and enable you to be very creative. The text-based tools work through commands in the text you write to create the diagrams automatically. They route the lines and place the shapes. It is a matter of preference. The graphical tools are perhaps easier to learn but the text-based tools are less labour intensive as all you have to do is type the words and the tool does the rest. The disadvantage of the text-based tool is the formatting. The diagram may not turn out as you would like it and requires patience to get right.

The graphical tools are best for paper presentation but are limiting for on-line application. It is this medium that the text-based tools score by allowing a diagram to extend ad infinitum. Some of the flowcharting tools have features for analysing processes in terms of costs, duration, resources, critical paths etc. This you cannot do directly with the other methods.

## Object linking

A flowchart can be produced to stand alone, be linked to a document or be linked to other flowcharts. By pasting charts in text documents you can either have object linking or no linking. The object linking allows you to update the chart from the document without loading the charting tool separately.

Advanced object linking will allow you to create several pages of diagrams and link individual shapes to other charts or documents. Visio and allCLEAR® both have this facility. This ability lends itself to the system design methodology explained in this book. The top page can be the system model and each shape can be linked to a chart and shapes in that chart linked to documents. The result is a hierarchy of files that can be accessed top down or bottom up. When you click on the shape you can load the associated chart. If the charts are pages of the same file then access is much faster than if the files are separate. If the link from a shape is to a document then access time is greater as the word processor is loaded into memory. All the links have to be made manually but you can give the charts names and navigate around the set of charts fairly easily.

The object linking method is suited to an on-line documentation system. By providing read-only access to the network version, integrity can be assured. With this method, the paper version will look identical to the electronic version. The principal disadvantage is access time and memory. You need a fast machine to run this application quickly. Waiting for applications to load will frustrate users unless it is instant.

## Hypertext features

A more advanced version of the object linked files is to develop the documents in Hypertext Mark-up Language (HTML), the kind of structure that is used on the World Wide Web. HTML links to other documents can be built into the text and all that is required is for you to click on the coloured text and the file is loaded. The links can be in any form so navigating through a hierarchy of charts can be done more easily. Access is much faster too as it is text-based.

A significant advantage of this method is the ability to search through the suite of files for key words or reference documents but it cannot give you relationships.

This method cannot be used in a paper environment because the structure is designed for linking. Many files may be created just to amplify a topic and so will not be stand-alone documents. The HTML method is therefore only suitable where everyone has access to a computer network.

One weakness of the HTML method is that it is a one-way path. You cannot navigate across the system or link to related documents unless you insert the links. If you have an object and want to see the relationships between that object and any other information you need a tool such as InfoMap. InfoMap is an information management tool designed to address the problems of managing information and knowledge.

# Deploying the product/service characteristics

### Identifying customer requirements

The purpose of the quality system is to provide the organization with the capability of satisfying its customers and is therefore worthless unless you have a clear understanding of what the customer requirements are. The specific requirements are not our concern as these will be implemented by applying the system. It is the range of requirements, needs and expectations that is important so that the system is designed with a certain capability. These will be Critical Success Factors (see Chapter 3).

### Risk analysis

A risk analysis should be carried out for each Critical Success Factor (CSF) to identify the processes where achieving success is at risk. The analysis should address the following:

- In which process are the requirements governing this factor defined? It may be defined in marketing, design or other processes.

- In which process are checks performed to verify that specified requirements have been achieved? It may be design verification, production test or other processes.

Between defining the requirements for success and measuring achievement, several things could go wrong. The next stage is to examine each process against the list of CSFs to establish:

- Whether any action or decision could adversely impact the CSF.

- What the nature of the impact would be.

- The effect this would have on the system.

- What controls have been built into the process to reduce or eliminate any adverse impact.

- What needs to be put in place to reduce or eliminate any adverse impact.

- The likelihood that the planned checks will detect an adverse impact on the CSF.

As the processes will be arranged in a hierarchy it is necessary to carry out the analysis at each level, otherwise some risks may be overlooked at the interfaces.

The results can be presented in tabular form, such as that illustrated in Table 5-3.

| **Process:** | Resource Maintenance | | **Parent Process:** | Resource Management | **Ref:** | RA07 |
|---|---|---|---|---|---|---|
| | | | | | **Date:** | 09/01/97 |
| **CSF** | **Action/ Decision** | **Nature of Impact** | **Effect on Service** | **Existing Controls** | **Improve- ments** | **Detection** |
| Security | Staff Transfer | Employee may take reprisals | Computer fraud, virus incursion, data interception, data corruption | Recover security passes Remove name rights from staff lists | Disable passwords and access | System alerts Surveillance |

Table 5-3 *Risk analysis*

The standard does not, in fact, require you to design an effective system. It does require the system to be reviewed to ensure its continuing effectiveness. However, if the system was not designed properly in the first place, the review may simply result in a series of minor improvements that are never ending and do not deal with the system as a whole. Many initiatives for quality improvement attack parts of the system but not the whole system. Improvement in processes is often made without considering the effects on other processes. This is certainly true with document changes where the effects of changing one document are not usually considered before authorizing the change. How, then, do you design an effective system? There are several techniques you can use. Failure Modes and Effect Analysis (FMEA), Fault Tree Analysis (FTA) and Theory Of Constraints (TOC) are but three. The FMEA is a bottom-up approach, the FTA a top-down approach and TOC a holistic approach.

One way of applying the FMEA technique to the quality system is to take each procedure objective and establish the probability of it not being achieved, the likely cause and effect on the system and the probability of the failure being detected by the downstream controls. The analysis may show up key activities for which there are no safeguards – activities that rely on one person doing something for which there are no checks that ensure the activity has been performed. The quality system is a collection of interrelated processes; therefore, by chasing the effect along the chain you may find single-point failures (parts of the system which affect the performance of the whole system). The FMEA approach is a bottom-up approach, looking at component failures and establishing their effect on the system. An alternative approach is to use a top-down approach such as Fault Tree Analysis to postulate system failure modes and establish which processes, procedures or activities are likely to cause such failures. The third method is relatively new. The Theory Of Constraints, developed by Eliyahu M. Goldratt in the 1980s, examines the system as an interconnection of processes and focuses on the one constraint that limits overall system performance. The theory is founded on the principle that if all parts are performing as well as they can, the system as a whole may not be. Each process links with others in a chain and therefore by improving one process you may degrade the performance of another. It looks for the weakest link in the chain of processes that produce organizational performance and seeks to eliminate it. Once eliminated, it looks for the next constraint on the system. Many of the constraints may not be physical. There may well be policy constraints that govern many of the actions and decisions being made. What may have to change is the policy for improvement in system performance to be achieved. In this way TOC is similar to FTA but goes beyond the physical boundaries of the system.

**Providing traceability**

The risk analysis may well result in many tables that are not user-friendly. A Critical Success Factor (CSF) deployment would solve this problem and provide at a glance an indication of the processes in which a particular CSF is addressed.

# Identifying documentation needs

## What to document

The policies and practices required to meet the quality policy and quality objectives need to be documented, otherwise the description of the system is incomplete. However, the degree of documentation varies from a simple statement of fact to details of how a specific activity is to be carried out. There has to be a limit on what you document. To document everything you do would be impractical and of little value. There are, however, several good reasons for documenting the organization's policies and practices:

- To encourage the people involved into thinking a problem through

- To convert solved problems into recorded knowledge so as to avoid having to solve them repeatedly

- To maximize company performance rather than departmental performance

- To cause people to act in a uniform way, thus making processes predictable

- To provide freedom for management and staff to maximize their contribution to the business

- To free the business from reliance on particular individuals for its effectiveness

- To provide legitimacy and authority for the actions and decisions needed

- To make responsibility clear and to create the conditions of self-control

- To provide co-ordination for inter-departmental action

- To enable the effects of potential process changes to be assessed

- To minimize variance and eliminate bottlenecks

- To improve communication and to provide consistency and predictability in carrying out repetitive tasks

- To provide training and reference material for new and existing staff

- To provide evidence to those concerned of your intentions

- To provide auditable criteria for execution against authorized practices

- To provide a basis for studying existing work practices and identifying opportunities for improvement

- To demonstrate after an incident the precautions that were taken or should have been taken to prevent it or minimize its occurrence

There are also some good reasons for not documenting certain policies and practices. If you can't predict the course of action or sequence of steps you need to take you can't write a procedure. You can't plan for unforeseen events and, as the unexpected will happen sooner or later, it would be wasteful of resources to produce procedures for such hypothetical situations. So, if you do not use statistical techniques, for instance, it is a waste of time writing a procedure that will not be used even though ISO 9000 may require one. You may be able to imagine a course of action but unless you are confident it will be necessary sooner or later, it is not worth the effort to write it down.

There are several other good reasons for not documenting policies and practices. Management may have no objection to doing many sensible things but may well resist declaring them as policy or prescribing them in published procedures. They may take this attitude for several reasons:

- Customers may use evidence of noncompliance no matter how trivial to terminate a contract or decline a tender.

- There may be many instances where the policy or procedure doesn't apply.

- They may wish to safeguard against over-zealous auditors or assessors.

- Managers may personally wish to choose the most appropriate action for given circumstances.

- They may wish to avoid overkill – doing more than is necessary.

- The practices may not have any effect on product or service quality.

- The practices rely on skills acquired by training where judgement is necessary to produce the desired result.

At school we are taught reading, writing and arithmetic so procedures should not attempt to define these functions. The procedures need only detail what would not be covered by education and training or is intuitive. A balance should be achieved between training and procedures. In order to provide training of consistent quality, it too should be documented in the form of training manuals, training aids and facilities. If you rely on training rather than employing documented procedures then you will need to show that you have control over the quality of training to a level that will ensure its effectiveness. We expect

**164** System design

staff to know how to do the various tasks that comprise their trade or profession: how to write, how to design, how to type, answer the telephone, how to paint, lay bricks, etc. You may feel it necessary to provide handbooks with useful tips on how to do these tasks more economically and effectively and you may also use such books to bridge gaps in education and training but these are not your procedures. The quality system has to be documented in your procedures, standards, guides or manuals.

## Types of documents

There are several types of documents used in work processes and each has a specific purpose. In general there are only three categories of information that need to be conveyed: requirements, plans and results. The requirements are what you have to achieve, the plans are the provisions made to achieve the requirements and the results are what you actually achieve. However, requirements, plans and results are conveyed in different ways. They attract different names dependent on the level in the system from which they originated and whether they are generated internally or externally to the quality system. The documents used to convey such information can be divided into four classes: policies, practices, derived documents and reference documents. Each document type can be assigned a two-letter code that defines a generic group of documents. The code can be used in document-identification conventions but alternative conventions can be used. A list of various codes is given in the box.

### *Policies*

These are the documents that define your intentions for meeting the obligations placed upon you. The document identification code for all policy statements could be PD, meaning a policy document.

These documents can be further divided.

> **Corporate policies**
> Statements that guide actions and decisions in order to achieve the goals of the enterprise. These tend to be general policies such as Environmental, Safety, Security, Financial, Investment, Expansion, Social, Personnel and Quality.

| | **Document Types** |
|---|---|
| AP | Activity Plan |
| AR | Activity Report |
| BF | Blank Form |
| BL | Blank Label |
| BP | Business Proposal |
| CA | Contractual Agreement |
| CC | Conformity Certificate |
| CP | Control Procedure |
| CR | Controlled Record |
| DN | Display Notice |
| ER | External Reference |
| GD | Guide |
| IA | Interface Agreements |
| IR | Internal Reference |
| JD | Job Description |
| OP | Operating Procedure |
| OR | Operations Record |
| PD | Policy Document |
| PM | Publicity Material |
| PO | Purchase Order |
| PS | Product Specification |
| ST | Standard |
| WI | Work Instruction |
| WO | Work Order |

## Business policies
Business policies are statements that guide actions and decisions in order to achieve specified business requirements. Examples are policies for Marketing, Sales, Pricing, Products and Services.

## Operational policies
Operational policies are statements that guide actions and decisions in order to achieve operational requirements. These tend to be detail policies such as Procurement, Inventory, Production, Servicing and a wide range of policies addressing some aspect of operations.

# *Practices*
These are statements that define the what, when, where, how, who and why of your operations.

## Control Procedures (CP)
Control procedures are documents that control work on a product or information as it passes through a process. They are generally multifunctional.

## Operating Procedures (OP)
Operating procedures are documents that prescribe how specific tasks are to be performed.

## Standards (ST)
Standards are documents that define acceptance criteria for judging the quality of an activity, a document, a product or a service. They change infrequently and can be applied to ranges of activities, products etc. There are international, national, industry and company standards.

## Guides (GD)
Guides are non-mandatory documents that are an aid to decision making and to conduct activities. There are other classifications such as manuals but these are a collection of documents such as guides, standards and procedures.

## Blank Forms (BF)
Blank forms are used to collect and transmit information for analysis or approval. The instructions on the blank form are an extension of the procedure that invokes its use. When a blank form is completed it becomes a record. It may also become a specification, procedure or other document depending on the type of information included.

## Blank Labels (BL)
Labels are documents that identify product status and are often disposed of when the status changes. The instructions on a blank label are an extension of the proce-

dure that invokes its use. When a blank label is completed it may become a record if containing results but labels are usually indicators only.

### Display Notices (DN)
Notices are documents that alert staff to regulations that must be followed, to precautions that must be taken, to dangers that exist. Notices are messages that transmit information contained in policies and procedures and are therefore an extension of the procedure that invokes their use.

### Job Descriptions (JD)
Job descriptions are documents that define the responsibility, authority and accountability of personnel. They are needed to assign responsibility for results so that everyone knows what is expected of them. Without job descriptions, there is a chance that some results will not be achieved simply because no one has been assigned. So that you are not constrained by any particular organization model you can use role descriptions (RD) instead.

## *Derived documents*
These are documents that are produced from implementing the policies and practices. The names of these documents vary significantly depending on the industry. There are two types of derived documents – prescriptive and descriptive. The prescriptive documents contain requirements and the descriptive documents contain results and intentions. Some common derived documents are:

### Activity Plans (AP)
Activity plans are prescriptive documents that define how specific requirements are to be achieved. There are many types of plans: Business Plans, Development Plans, Reliability Plans, Maintenance Plans, Quality Plans, Production Plans etc. They all define specific measures or provisions to meet certain requirements and are in general time-limited but do not have to be. A disaster recovery plan for a data centre, for instance, is implemented whenever there is a data storage failure. An emergency plan for a chemical plant is implemented when there is a chemical emergency. Contingency plans are implemented when the routine plans have not worked.

### Product Specifications (PS)
Product specifications are prescriptive documents that define acceptance criteria for specific products or services. There are many types of specifications: product specifications, service delivery specifications, test specifications, process specifications etc. Engineering drawings are specifications in diagrammatic form.

### Work Instructions (WI)
Work instructions are prescriptive documents that define specific tasks required and cause them to be accomplished, detailing when they are to commence and be completed. Instructions can be short, such as a note for the Secretary to type a document

by Monday morning, or long, such as an assembly instruction that gives details of how to assemble a piece of equipment.

### Business Proposals (BP)
Business proposals are descriptive documents that describe a solution to a requirement offered for consideration by a customer or upper manager.

### Advertising Literature (AL)
Advertising literature comprises descriptive documents that describe the products and services offered by a company. They are not usually binding on the company and clauses can be inserted to give the company the right to change specifications without notice.

### Conformity Certificates (CC)
Conformity certificates are descriptive documents that testify that certain requirements have been met. Examples are test certificates, release certificates, certificates of conformity, training certificates.

### Activity Reports (AR)
Activity reports are descriptive documents that describe the results of some activity or project. Examples are progress reports, annual reports, audit reports, test reports. Generally, reports contain conclusions from results obtained from records.

### Operations Records (OR)
Operations records are descriptive documents that contain factual information in a raw state. They may be lists, tabulated data or graphical data. The information may also be contained on a form or label. These records are transient and change quickly (such as minutes of meetings). Their development, issue, revision etc. does not need to be controlled.

### Controlled Records (CR)
Controlled records are descriptive documents that contain objective evidence of the achievement of specified requirements. Quality records, environmental records and safety records are a special form of record and result from measurements, assessments or evaluations taken in a process. The information may also be contained on a form or label. The information on these records does need to be controlled as it will be used to demonstrate compliance with requirements, defend a prosecution or demonstrate system effectiveness.

### Internal Reference Documents (IR)
Internal reference documents contain data that is to be used in conjunction with the policies and practices or other documents. They are not instructional nor do they define criteria for making decisions. Often they consist of lists of names, addresses and other data.

## External reference documents (ER)

These are documents that are used for reference purposes in implementing policies, practices and derived documents. There are several types of reference documents, many of which can be classed as policies, practices or derived documents. The feature that makes them external reference documents is that they are produced outside the quality system and are used as inputs or guides in the implementation of the system. One reason for separating the external reference documents from the internal reference documents is to enable different controls to be applied. You do not normally have the authority to change external documents. One can control only their use, condition and internal distribution.

The following are examples.

### Contractual Agreements
Contracts are documents issued by the customer that specify requirements that the company has accepted for supplying specific products and services over a specified duration and for a specific price.

### Interface Agreements
Interface agreements are documents that define the agreed level of service between two parts of the same company where each part has its own quality system. Although produced within the organization and not within the quality system, they are reference documents because to change them requires the agreement of parties outside the system.

### Purchase Orders
Purchase orders are documents issued by the customer for specific supplies and services at a price stipulated by the supplier. Purchase orders issued by the organization are internal documents.

### Work Orders
Work orders are instructions from customers for specified work to be executed. They may be issued under the authority of a contract or purchase order. Work orders issued by the organization are internal documents.

### Standards
National, international, industry and customer standards are documents that prescribe general requirements that are to be met. Standards issued by the organization are internal documents.

### Drawings and Specifications
Customer drawings and specifications that specify the acceptance criteria for the products and services to be supplied.

## Processes and procedures

Process and procedures are quite different in nature. A process is not a procedure. Processes include procedures as the methods employed to convert inputs into outputs whereas the process includes everything needed to meet the exit conditions. Processes include the materials, equipment, environment, personnel and measurements to transform inputs into outputs of added value. A process is a chain of operations (with or without hold points) that achieves a measurable objective. The methods used in a process may be defined in one or more procedures. A procedure is a sequence of uninterrupted tasks that have defined entry and exit conditions. A procedure may end without achieving a measurable objective.

## Procedures and documents

Procedures have entry and exit points. The documents that describe the procedures may contain one or more procedures. You do not need to produce separate documents for each procedure. It is a matter of judgement. If a sequence is short then its inclusion in the parent document is more practical than producing separate documents just for the sake of it. If the procedure will be used by the same people as the parent document then creating a separate document will just produce more paper without any added value. Look at the needs for referencing, maintenance and distribution when choosing whether to produce separate documents. When identifying documentation needs you need to establish the boundaries of the procedure documents and these may change depending on the following criteria:

- Common routines should be placed in separate documents to ease referencing, maintenance and distribution. For example, problem reporting may be common to several processes. Rather than describe the procedure several times, it is more efficient if described once in a separate procedure.

- Changes in continuity create suitable exit points for a document. For example, in the purchasing process the tasks performed up to placement of the order should be described in one document and the tasks performed after the goods have been received in another document. In this way you can exit the procedure and move on to another task. By creating separate documents in this case, you can use the same receiving procedure for dealing with goods that did not originate from a purchase order without having to create a new procedure. Inter-site transfers would be one example.

- Changes in responsibility create suitable exit points for a separate document particularly for operating procedures. In defining a work process, several tasks may be performed sequentially by one function and then several more by another function. Drawing the boundary around the tasks with the same responsibility creates suitable entry and exit conditions for a documented procedure.

- Repetitive sequences placed in separate documents ease document maintenance as they can be referenced when required rather than repeated. Unique sequences placed in separate documents ease referencing and distribution as they will be used only under certain conditions. An example is a disaster recovery procedure. It is not cycled every day, in fact only when there is a disaster so should be documented separately.

## Documentation analysis

There are several approaches you can take to identify the documents needed in your management system. You can:

- Search the relevant standards for references to documents.

- Respond to the requirements of the standards with operational policies and then identify where documents would be needed to implement them.

- Chart the processes and identify where documents are needed as inputs to a process, where documents are needed for reference purposes in executing the process and where documents are produced as outputs of the process.

Let us take one example from ISO 9001 to illustrate the differences.

In clause 4.1.1 in ISO 9001 the requirements are stated as follows:

> The supplier's management with executive responsibility **shall** define and document its policy for quality, including objectives for quality, and its commitment to quality. The quality policy **shall** be relevant to the supplier's organizational goals and the expectations and needs of its customers. The supplier **shall** ensure that this policy is understood, implemented and maintained at all levels in the organization.

Notice that there are three separate requirements but only one that explicitly calls for documentation.

### *Documents derived from the standard*
There is only one document specifically required in clause 4.1.1. The other requirements can be demonstrated by deed rather than documentary evidence. This represents the minimum requirement but when compared with the second example it will become clear that other documents (or pieces of information) are in fact needed to run the business effectively.

System design 171

| Requirement | Response | Documentation |
|---|---|---|
| *The supplier's management with executive responsibility **shall** define and document its policy for quality including objectives for quality, and its commitment to quality.* | The company's policy for quality including objectives for quality, and its commitment to quality is defined in the Quality Policy. | Quality Policy |

## *Documents derived from operational policies*

In the following table, each requirement from clause 4.1.1 of ISO 9001 has been dissected and a response provided in the form of an operational policy. These policies indicate a need for a document of some type to be provided and these are identified in the third column.

| Requirement | Response | Documentation |
|---|---|---|
| *The supplier's management with executive responsibility **shall** define and document its policy for quality.* | The company's philosophy and hence commitment to quality is declared in a Policy Statement signed by the Managing Director. | Quality Policy |
| *The supplier's management with executive responsibility **shall** define and document its objectives for quality.* | In order to maintain established quality standards and facilitate continuous improvement in the quality of the company's products and services, quality objectives shall be established each year and a programme implemented to achieve them. | Quality Objectives<br>Quality Improvement Programme |
| *The supplier's management with executive responsibility **shall** define and document its commitment to quality.* | The Quality Policy Statement, the Quality Improvement Programme and the resources to carry it through shall be authorized by the Managing Director. | Quality Policy<br>Quality Objectives<br>Document Approval Records |

| Requirement | Response | Documentation |
|---|---|---|
| *The quality policy **shall** be relevant to the supplier's organizational goals.* | The quality policy shall be developed in conjunction with organizational goals that have been defined in the 1, 3 & 5 year Business Plan. | Business Plan |
| *The quality policy **shall** be relevant to the expectations and needs of its customers.* | The quality policy shall be periodically reviewed by the management team against the market analysis of current customer expectations and needs and amended as appropriate. | Market Analysis Report |
| *The supplier **shall** ensure that this policy is maintained at all levels in the organization.* | The policies for enabling this philosophy to be maintained throughout the company's operations shall be documented in a Quality Manual, approved by the management and copies made available to all staff. | Quality Manual |
| *The supplier **shall** ensure that this policy is understood at all levels in the organization.* | These policies shall be developed in conjunction with the staff who will implement them and only published once they have been agreed and understood. Managers shall coach their staff to ensure understanding and provide help and advise in the implementation of the published policies and procedures. | Staff briefing notes<br>Staff appraisal records |
| *The supplier **shall** ensure that this policy is implemented at all levels in the organization.* | Procedures and standards shall be developed to implement the agreed policies, referenced within the Quality Manual and subject to a comprehensive programme of independent audits. | Procedures and standards<br>Audit reports |

## Documents derived from charting the process

It is not usual for organizations to chart the process for generating the quality policy. However, the thought processes involved are real and the quality policy clause of ISO 9001 is not all that easy to chart as a series of processes because it is not a routine exercised every day. The quality policy and quality objectives emanate from the Strategic Planning task and result in the creation or improvement to the quality system. A typical Business Management Process is illustrated in Figure 5.34. On the diagram have been

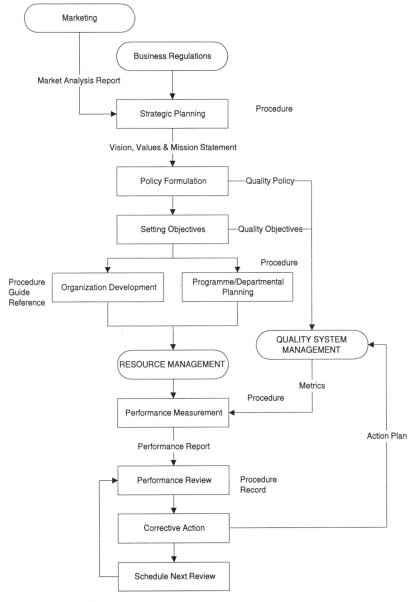

Figure 5.34 *Business Management Process*

placed the titles of various documents that are either inputs or outputs of the process or methods used in the process. This method results in a different set of documentation than the other methods.

If we analyse the business management process applying the rules defined previously we can identify the following documentation needs.

| | |
|---|---|
| **Business Management Policy** | This will be part of the Policy Manual and will establish the requirements governing the process. |
| **Business Management Procedure** | This will be a control procedure (CP) because it documents a key process and passes information through the organization. |
| **Strategic Planning Procedure** | This will be an operating procedure (OP) because it will be executed by one group of staff, the executive. Details could be included in the control procedure if it is felt that a separate document is not warranted. |
| **Market Analysis Report** | This is an output of the Marketing Process and will be an activity report (AR). |
| **The Vision, Values and Mission Statements** | These will be policy documents (PD) and will probably be included in the Policy Manual. They may also be converted into display notices (DN). |
| **Quality Policy** | This will be a PD but may also be a DN. |
| **Quality Objectives** | This will be a PD but may also be a DN although quality objectives would have to be changed more frequently than the quality policy because the objectives hopefully would be achieved at a certain point and new objectives defined. |
| **Programme/ Departmental Planning Procedure** | This will be an operating procedure as it will define a sequence of steps for a manager to follow. There may be approval loops that cause the plan to pass through the organization and back again. |
| **Organization Development Procedure** | This will be an operating procedure that defines how appointments are to be made, jobs defined and departmental structure displayed. |

| | |
|---|---|
| **Organization Chart** | This will be a reference document depicting the structure of the company and its component parts. |
| **Organization Development Guide** | This will guide managers in choosing job titles, making appointments, defining responsibilities etc. |
| **Metrics** | These will be transmitted on forms, graphs or other means as operational records (OR). |
| **Performance Measurement** | This will be an operating procedure (OP) because the task will probably be performed by one person or group. |
| **Performance Report** | This will be an controlled record (CR) rather than an activity report because it will contain the objective evidence of the performance resulting from the performance measurement task. |
| **Performance Review** | This will be an operating procedure (OP) because it will cover the review of the performance report by the management team. |
| **Action Plan** | This will contain the results of the review and detail the actions required to bring performance in line with requirements. The plan will also show the date when the next review has been scheduled. |

These documents describe the Business Management Process. In this example, the Business Process of Business Management results in one control procedure. In other cases, the Business Process Chart will need to be divided into subsystems and each of these covered by one or more control procedures. There are no hard and fast rules. It is largely a matter of judgement as to where the boundaries fall and what goes into which documents.

## Requirements of ISO 9000

ISO 9001 defines several requirements that apply to documentation. In summary the standard requires:

- A system to be established, documented and maintained

- Quality system procedures to be prepared that are consistent with the standard and the quality policy

The standard also requires several documents to be produced, a list of which is given in Annex A. These are the minimum requirements for assuring the attainment of quality but of course are not the only documents you will need. If you examine ISO 9004-1 or ISO 9004-2 there are many other documents that comprise a quality system[5]. A particular quality system will include other documents. The documents needed should always be relevant to the business and in so doing, some documents required of ISO 9001 may not be applicable and some of those recommended in ISO 9004 may not be appropriate. It is the business that dictates what documents are needed not the standard. It is therefore necessary to interpret the requirements of the standard in appropriate ways and not create documents just because there is a requirement. ISO 9000 is a model and not a prescription, although it is a requirement to be interpreted as appropriate to the business being examined.

## Structuring the documentation

Documentation is but a description of the business and how it operates. There are, however, levels of documentation – documents that summarize the system and others that specify in increasing detail the what, why, how, where and who. The structure can be depicted as a pyramid of documentation (as illustrated in Figure 5.35). In the diagram, the term *output documents* is used to represent all the derived documents. The model is intended to illustrate that at each level there are more and more documents that support the level above. Whilst in theory this may be valid, in practice the reality is far different for there is no reason why:

- A record or output document should not be a level below a policy

- A control procedure should not reference another control procedure

- A form should not be derived from an operating procedure

- A control procedure or a policy or an operating procedure should not be derived from an output document

In reality, documentation should have a hierarchical structure rather than a pyramid structure.

ISO 9001 requires the documentation structure to be described in the Quality Manual. This could be accomplished by a pyramid diagram as above but, as stated previously, this is not strictly accurate. A more accurate description is provided by explaining the development methodology. A simple description might be as illustrated in the box on page 182. You can then define each type of document referred to. If you have a paper

---

[5] See *ISO 9000 Quality Systems Handbook* for a comprehensive list.

## System design

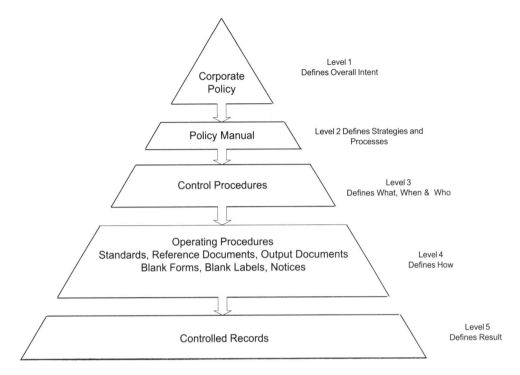

Figure 5.35 *Documentation hierarchy*

system, then it may be necessary to describe the volumes that contain the documents. One solution is to provide three manuals as follows:

- Quality Manual, containing the quality policy and operating policies

- Procedures Manual, containing the control and operating procedures

- Support Documentation Manual, containing the forms, standards and other documents

The problem with this solution is that it forces documentation into pigeon holes that may be inappropriate. Departments often prefer departmental manuals containing their standard practices. The derived documents do not easily fit into categories and are often divided by product, project or contract.

It is therefore more appropriate to separate documents into the four categories of policies, practices, derived documents and reference documents. The policies and practices can, however, be structured as a Policy Manual and a Procedures Manual with the remainder being subject to local policy on a case-by-case basis.

## 178 System design

> ### DOCUMENTATION STRUCTURE POLICY
>
> - The quality system documentation consists of policies, practices, derived documents and reference documents.
>
> - Policies define the boundary conditions for carrying out actions and decisions. These exist at three levels: the enterprise, business and operational levels. The enterprise policies with respect to quality are defined in a corporate quality policy statement. All the other policies are defined in a Policy Manual, control or operating procedures.
>
> - Practices implement policies and are described by flowcharts, control procedures, operating procedures, standards and guides. Control procedures implement policies and operating procedures implement tasks specified in control procedures.
>
> - Both internal and external reference documents are employed as a source of information that is used in the implementation of the documented practices
>
> - Evidence of activities specified in control or operating procedures and the results of achievements are documented in the form of controlled records.
>
> - All documents are traceable through the documentation hierarchy to the policies contained in the Policy Manual.

## Practical document identification

Three techniques for identifying documentation needs have been described above and the process-charting technique is the most practical as it will produce a documentation list matched to the business needs. However, charting a process and guessing what documents are needed is no better than dragging the documentation requirements out of ISO 9001. Here follows a more practical method.

1. Chart processes down to task level.

2. If the tasks cross functional boundaries then you have identified the need for a control procedure.

3. Produce the control procedure following the tasks identified on the chart.

4   For the first task on the chart determine what form the input takes. If the input is in the form of *information* you will have identified a derived document from the interfacing process.

5   Determine what activities are required to carry out the task. If there are many sequential activities, you have identified an operating procedure.

6   If the task is one that will generate information, determine what form the derived documents will take. This will identify plans, specifications, reports or derivations of the same.

7   If the task is an acceptance decision for which criteria needs to be specified, you will have identified the need for a standard.

8   If the task is a moving-type task you may have identified a requirement for labels and notices.

9   If the task is one where personal or equipment safety is at risk, you may have identified a requirement for standards, guides and notices.

10  If the task is a decision for which there may be many possible options and for which some consistency is required, you will have identified the need for a guide.

11  If in the execution of the task, *reference material* is to be consulted then you have identified the need for a reference document.

12  Determine whether it is necessary to maintain a record of the results of the decision or action to feed other processes. Should such a record be needed you will have identified a controlled record.

13  For the final task in the flowchart, determine what form the output takes and if it is *information* determine the type of derived document it is.

14  If the output of the process is product, then you will have identified a requirement for labels and routing instructions.

The important thing to remember when identifying documentation is that the process should be systematic otherwise a degree of randomness will creep in and cause documents to be generated for which there is no requirement. The requirement for the document derives from the description of the policy, task or activity which causes it to be generated. Here are some examples.

180  System design

- The plan shall be prepared in accordance with procedure OP23.

- The results shall be recorded on form BF45.

- The product shall be inspected using standard ST 04.

- In deciding which course of action to take, consideration shall be given to the guidance provided in Guide GD14.

- An acceptance label (BL09) shall be attached to those products that have passed the specified tests.

- When installing replacement blades in the machine, the safety precautions defined in DN06 shall be observed.

Documentation identification is therefore a progressive process. You chart the process, identify a document or two, chart another process, identify a few more documents, write a procedure and identify many more, produce the documents you identified and identify many more documents, and so on until all have been identified and produced.

## Summary

In this chapter we examined several methods for designing quality systems and presented a case for using the business process methodology as a viable alternative that provides an enduring solution to quality system design. Key messages from this chapter are as follows:

- Quality systems are not a collection of procedures but an interconnection of processes that deliver desired outputs.

- To cause anything to happen there has to be a chain reaction.

- What turns a collection of components into a system is the way they are connected and energized.

- When we design systems, therefore, we start not with a set of components to see what we can make from them but with a diagram showing the primary functions and how they are interconnected.

- In designing anything there has to be a requirement – a definition of what the product or service has to achieve.

- If you document what the standard requires you may end up with a 'bolt on' system where the documents are generally descriptive of what you do but are not actually used.

- If you document what you do you might be making the assumption that what you are doing is OK (it meets the standard) when in fact this could be far from the truth.

- The problem with placing all your trust in staff to control the outcomes of the business is that you never know what it is they do that you are relying on until they no longer do it.

- A business-process approach creates a coherent system that matches business needs, meets the standard, causes conformity and prevents nonconformity through a chain of interconnected processes.

- There are four levels in the hierarchy of the Quality System Design. The first level is the system, which comprises business processes. Business processes comprise work processes. Work processes comprise tasks. Tasks comprise activities. This final division forms the basis of procedure models. Sometimes, we need to group business processes together into subsystems to permit progressive decomposition.

- The business process methodology doesn't start with the development of procedures but with the definition of the business processes and developing successive levels of flowcharts to a level where specific tasks can be defined and the need for documentation identified. Some functions and requirements may apply to only one process whereas others may apply to many processes.

- If you don't identify documentation systematically in a progressive manner, a degree of randomness will creep in and cause documents to be generated unnecessarily.

## System design task list

1. Produce a system requirement specification.
2. Create context diagram.
3. Create system model.
4. Chart business processes.
5. Chart the work processes that comprise the business processes.
6. Chart the tasks that comprise the work processes.
7. Deploy the regulations to the process models.
8. Deploy functions to tasks.
9. Perform risk analysis against the critical success factors.
10. Add controls to meet relevant requirements.
11. Add controls to reduce or eliminate adverse impact on critical success factors.
12. Identify types of documents needed to transmit input requirements, perform work and supply process outputs.
13. Perform document reduction to identify minimum separate documents.
14. Create document register.
15. Match existing documents against identified needs and update document register.
16. Define the development requirements for the listed documents.

## System design questionnaire

1. Have you decided what form of system you require: paper-based or electronic?

2. Have you identified all external business and system interfaces?

3. Have you identified the core business processes?

4. Have you charted the business processes and associated work processes?

5. Have you defined the business and work processes and set the boundary conditions?

6. Do all the inputs and outputs on the work process charts link with the corresponding interfaces on the associated business process chart?

7. Do the charts match your process descriptions?

8. Have you taken account of optional routes for product and information to follow as it passes through the process?

9. Have you given names to the processes?

10. Do the names reflect the function of the process rather than the organization which executes it?

11. Have you identified the documentation or product which passes between tasks on your flowcharts?

12. Have you checked that all functions have been represented on your Function Deployment Matrix?

13. Have you identified the corresponding tasks on the flow charts which will implement the requirements in your Requirement Deployment Matrix?

14. Have you added all the necessary controls and responded to the relevant requirements identified in the matrix?

15. Have you completed your risk analysis and put in place the necessary controls to minimize the risks as product or information passes through the processes?

16. Have you identified the principle control procedures required and charted the corresponding flowcharts?

17. Have the flowcharts been agreed by the process owners and development teams?

# Annex A
# Documents required by ISO 9001

The following table identifies the types of documents required by ISO 9001 and the business process in which they would be generated. A more comprehensive list of subjects is given in my *ISO 9000 Quality Systems Handbook*, but this list will assist in identifying the corresponding company documents that respond to the requirement.

| No | Subject | ISO Ref | Business Process |
|---|---|---|---|
| 1. | Quality policy | 4.1.1 | Business Management |
| 2. | Quality objectives | 4.1.1 | Business Management |
| 3. | Responsibility and authority of key personnel | 4.1.2.1 | Business Management |
| 4. | Interrelationship of personnel | 4.1.2.1 | Business Management |
| 5. | Management review records | 4.1.3 | Quality System Management |
| 6. | Quality manual | 4.2.1 | Quality System Management |
| 7. | Quality planning | 4.2.3 | Product/Service Planning |
| 8. | Contract review procedure | 4.3.1 | Marketing |
| 9. | Contract requirements | 4.3.2 | Marketing |
| 10. | Amendments to contract | 4.3.3 | Marketing |
| 11. | Contract review records | 4.3.4 | Marketing |
| 12. | Design control procedures | 4.4.1 | Product/Service Design |
| 13. | Design verification procedures | 4.4.1 | Product/Service Design |
| 14. | Design and development plans | 4.4.2 | Product/Service Design |
| 15. | Organization interfaces | 4.4.3 | Product/Service Design |
| 16. | Technical interfaces | 4.4.3 | Product/Service Design |
| 17. | Design input requirements | 4.4.4 | Product/Service Design |
| 18. | Design output documents | 4.4.5 | Product/Service Design |
| 19. | Design review records | 4.4.6 | Product/Service Design |
| 20. | Design verification measures | 4.4.7 | Product/Service Design |
| 21. | Design changes | 4.4.9 | Product/Service Design |
| 22. | Document control procedures | 4.5.1 | Quality System Management |
| 23. | Master list of documents | 4.5.2 | Quality System Management |
| 24. | Purchasing procedures | 4.6.1 | Resource Acquisition |
| 25. | Subcontractor control plan | 4.6.2 | Resource Acquisition |
| 26. | Subcontractor audit reports | 4.6.2 | Resource Acquisition |
| 27. | Records of acceptable subcontractors | 4.6.2 | Resource Acquisition |
| 28. | Purchasing documents | 4.6.3 | Resource Acquisition |

| No | Subject | ISO Ref | Business Process |
|---|---|---|---|
| 29. | Customer supplied product control procedures | 4.7 | Resource Acquisition, Production |
| 30. | Customer supplied product records | 4.7 | Resource Acquisition |
| 31. | Product identification procedures | 4.8 | Product/Service Design, Production |
| 32. | Traceability procedures | 4.8 | Production |
| 33. | Identification records | 4.8 | Production |
| 34. | Production plans | 4.9 | Production |
| 35. | Installation plans | 4.9 | Installation |
| 36. | Servicing plans | 4.9 | Servicing |
| 37. | Production procedures | 4.9.1a | Production |
| 38. | Installation procedures | 4.9.1a | Installation |
| 39. | Servicing procedures | 4.9.1.a | Servicing |
| 40. | Workmanship criteria | 4.9.1f | Production, Installation, Servicing |
| 41. | Process qualification requirements | 4.9 | Production |
| 42. | Process qualification records | 4.9 | Production |
| 43. | Inspection and test procedures | 4.10.2 | Product Planning |
| 44. | Inspection and test specifications | 4.10.1 | Product Planning |
| 45. | Premature release records | 4.10.2.3 | Resource Acquisition |
| 46. | Inspection and test records | 4.10.5 | Production, Installation, Servicing |
| 47. | Inspection measuring and test equipment | 4.11.1 | Resource Maintenance, Control Procedures |
| 48. | IMTE calibration procedures | 4.11.1 | Resource Maintenance |
| 49. | IMTE maintenance procedures | 4.11.1 | Resource Maintenance |
| 50. | Test devices records | 4.11.1 | Resource Maintenance |
| 51. | IMTE technical data | 4.11.1 | Resource Maintenance |
| 52. | Calibration instructions | 4.11.2 | Resource Maintenance |
| 53. | Calibration records | 4.11.2 | Resource Maintenance |
| 54. | Out-of-calibration assessment reports | 4.11.2 | Resource Maintenance |
| 55. | Nonconforming material control procedures | 4.13.1 | Resource Acquisition, Production, Installation, Servicing |
| 56. | Nonconformance reports | 4.13.2 | Resource Acquisition, Production, Installation, Servicing |

| No  | Subject | ISO Ref | Business Process |
|-----|---------|---------|------------------|
| 57. | Corrective action procedure | 4.14.1 | Quality System Management |
| 58. | Preventive action procedure | 4.14.2 | Quality System Management |
| 59. | Customer complaints procedure | 4.14.2a | Quality System Management |
| 60. | Procedures change records | 4.14.1 | Quality System Management |
| 61. | Investigation reports | 4.14.b | Quality System Management |
| 62. | Handling procedures | 4.15.1 | Resource Acquisition, Production, Installation, Servicing |
| 63. | Storage procedures | 4.15.1 | Resource Acquisition |
| 64. | Packing procedures | 4.15.1 | Order Fulfilment |
| 65. | Preservation procedures | 4.15.1 | Resource Acquisition, Production, Installation, Servicing |
| 66. | Delivery procedures | 4.15.1 | Order Fulfilment |
| 67. | Quality records control procedure | 4.16 | Quality System Management |
| 68. | Internal quality audits procedure | 4.17 | Quality System Management |
| 69. | Audit reports | 4.17 | Quality System Management |
| 70. | Training needs procedure | 4.18 | Resource Maintenance |
| 71. | Training records | 4.18 | Resource Maintenance |
| 72. | Servicing procedures | 4.19 | Servicing |
| 73. | Service reports | 4.19 | Servicing |
| 74. | Statistical techniques control procedure | 4.20.2 | Resource Acquisition, Production, Installation, Servicing |

## Chapter 6

# Documentation development

## Document requirements

By applying the principles covered in Chapter 5 you will be able to identify those policies and practices which need to be documented. For every document identified there is a known need but it is necessary to define other characteristics so that the document produced will satisfy that need. You could rely on the definitions of the generic document types to convey the needs but this is often insufficient. Once documents are identified the next stage is to assign owners and select authors. Without some idea of what they have to write, the chances are that the authors will not produce a document that fulfils the need. An instruction to a person such as 'Please produce a procedure for purchasing' may get the response 'Purchasing what?' After that question, another might be: 'What do I include, what do I exclude?' A flow diagram should give the boundary conditions, so the relevant flow diagram should be given with the instruction. You will probably meet with the nominated author to discuss what is required but unless you provide a written statement the chances are that some aspects may be overlooked. If the author is some distance away so that conversation is mainly by phone, there is much greater need for clear, written communication.

As a minimum, the requirement should include the following:

- Document title
- Purpose and scope defining the boundary conditions
- Identity of parent process
- The flow diagram identifying the process, task or activity requiring documentation
- The constraints

You could define documentation standards which specify the minimum requirements for certain documents.

## Development planning

### The documentation development plan

As documentation needs are identified, a Document Development Plan should be created. For each document the plan should identify the following:

- Document reference
- Document title
- Parent process
- Document owner
- Document author
- Development status

### Selecting owners

The document owners are nominated by the process owners. They may be one and the same but, with a process that covers many different functions, document owners may be nominated for each function. The document owner has the same responsibilities as the process owner, except that they are limited to specific documents (see Chapter 4).

### Selecting authors

Criteria for selecting authors is given in Chapter 4. The author may not be the document owner. The author may be someone who has the knowledge and skill to write the document and the time available. Whilst the document owner could write the document, he/she may not have the time available if of managerial status.

### Document status

A column for document status is useful for keeping track of your progress. There are five indicators that can be used:

**Assigned** The document requirement has been assigned to the nominated author.
**Draft** A draft has been produced.
**Reviewed** Comments have been passed back to the author.
**Approved** The document has received the approval of the relevant authorities.
**Withdrawn** The need for the document has been eliminated.

## Documentation methods

### Options

There are several suitable methods that can be used for documenting your practices. All have some merit and it is a matter of culture and complexity that will dictate which method is appropriate.

Control procedures may be represented by flowcharts and forms – with notes on the charts and form-filling instructions on the back of the forms. For simple quality systems this method has many advantages: it is concise, easy to produce, easy to understand and easy to use. The main vehicle for transmitting information is the form, both in simple and in complex quality systems. Information passes from one department to another by use of forms so it makes sense to devise your control procedure around the forms you use. You probably won't be able to do this for operating procedures that require instruction (for example, on how to operate a piece of equipment, how to perform a test or how to carry out a calibration).

### *Text-based approach*
A text-based approach is suitable for standards, guides and operating procedures either because there is no work flow or because the work flow is sequential without diversions, optional routes or transitions between functions.

The headings in a text-based document should follow in a logical sequence in order for the reader to follow the logic of the instructions. If the headings are in a random order, the reader may get confused.

The text itself should provide concise, clear, complete and unambiguous instructions regarding the subject matter, including any references to other documents, equipment etc. that are needed to complete the task.

### *Flowchart-based documents*
Flowchart-based documents are only suitable for procedures although guides can sometimes use flowcharts for diagnostic routines. There are two types of flowchart that can be used: a process flowchart or a cross-functional chart (examples of which are given in Chapter 4).

## Combination

There are three ways in which you can combine charts with text.

### Combination charts
If all the information required can be placed inside the boxes on the chart then no further text will be necessary. An example of this type of procedure is given in Annex A.

### Flowchart-based with notes
Another way of combining charts with text is to make the flowchart the principle instruction and add notes to amplify the words in the boxes. The notes are given a reference code and pages of notes are appended to the chart explaining the detail. Depending on the amount of detail, this type of layout can be as comprehensive as the text-based with flowchart or as brief as the combination charts.

### Text-based with flowchart
The most common method of combining charts with text is to explain the flowchart through text, one box at a time so that the headings in the text follow the chart. In this way the chart acts as an aid to understanding the flow and leaves the chart free from detail so that responsibilities, reference documents and clear instructions can be given without the constraint of a chart. The charts help to illustrate aspects that are difficult to describe by text and the text is used to explain aspects that are difficult to show on a chart. An example of this procedure is given in Annex A at the end of Chapter 7.

## Producing effective documentation

Effective documentation leaves nothing to chance, no assumptions, no vagueness and clearly explains what is required to be done, who is to do it, the tools required, the inputs/outputs and routing instructions.

Even when you think the author understands what is required, the result can be far from what you expect. A task on the flowchart may be 'Update Records' and the written response by the author is to state that the manager will update the records. What records? Where are they kept? What details are to be updated? These are typical questions that vague statements provoke. If the procedure is to cause things to happen, as it should, you cannot get away with vague statements. Statements have to be specific, otherwise you are relying on prior knowledge which has to be passed down by the managers. What happens if the manager is not there to give instruction – how might the person charged with the task perform it? It is likely that some records will be updated correctly because the person uses common sense or intuition, but he/she is unlikely to be clairvoyant and, therefore, may not know of all the records that are to be updated. The records must be identified in order to prevent problems later. This is the key to effective procedures. You should anticipate what could go wrong, what might be left undone and specify the details necessary to avoid people making mistakes, neglecting to do important tasks etc.

When you start to document the policies and practices you need to get into the right frame of mind. Think what you are about to do. You could, for instance, be thinking of writing any one of the following:

- War and Peace
- A memo to the boss
- A short story
- An overview
- A business proposal

In fact, you are writing none of these. The result will inevitably be the result of your education and commensurate with your current position. If you are a department manager writing the documents, you may be more inclined to brevity and may make assumptions about what people know. You may only write what you think is important. If you are an operator, you may be more inclined to write in more detail but only in the area in which you work. You may not know about the interfaces and may not think to add details of where things come from, where they go to etc. You may take it for granted that everyone who works there knows.

If you are a consultant you may be inclined to follow the standard or be prescriptive – over-specifying the requirements, thereby making a straitjacket for the implementers. Consultants have also been known to be inventive where no practice exists and not gain agreement from the users before the procedures are issued. If you are a computer programmer you may write very detailed procedures concerning what is done but very little about how it is done. You may specify your entry and exit conditions, the feedback loops and responsibilities but may not think about the controls applied to the tools you use, the environment in which you develop the software, etc.

It is therefore not easy to produce effective documentation. You may have to escape from your 'comfort zone' to do so. Firstly, you need to know your readers (see panel). If you were to write a document on the same subject for each of these readers, the style and content would be different.

**KNOW YOUR READER**

- Users
- Managers
- Consultants
- Auditors
- Customers
- Trainees

> ## THE RED TEAM
>
> An organization had produced several manuals of common procedures that applied to all the functions of the business. The procedures contained prescriptive and descriptive material including the forms. Each department only had to produce procedures where their activities were at variance with the common procedures. These common procedures gave instruction as well as reasons why the tasks should be carried out and were therefore very detailed. When a new Managing Director was appointed and saw the volumes of procedures spread out in his bookcase, he instructed the Quality Director to reduce the documentation by half. A Red Team was formed and commenced the document-reduction task. It was called a Red Team because they used red pens to strike through anything that they felt was not essential.
>
> Months later the new slim documents emerged and the Managing Director was pleased. Several months after that, though, each department having examined the new slim volumes started to produce more documents defining what they did, filling the gaps that had been created by removing explanatory text from the common procedures. And so the overall weight of the volumes was not reduced. Rather than promoting harmonization, the MD had caused the devolution of practices. Where there was one common practice there were now several. Increasing the complexity of the system increased the effort to audit it.

However, we are not doing this. We could but it would make the system impossible to maintain. We need one document for all readers. We could adopt an approach whereby the procedures are prescriptive and brief with separate guides to enlighten the uninitiated. The problem with this approach is the difficulty of determining what goes into the procedure and what goes into the guide.

When you have produced a document, conduct the following tests after each statement as appropriate.

- So what?

- What's next?

- Where did that come?

- What do we do with that?
- How is that done?
- Who does this?
- Did we use this term on the last occasion in this context?
- Is this task or decision within the scope of the role we have defined?
- Do we really intend that this happens every time?
- Is the approval level too high?

If the test reveals no sensible result the text needs to be revised.

## Documentation tools

The tools used to create the flowcharts can also be used to link to documents and, with some tools, the instructions explaining how an activity is to be performed can be linked to a separate file or pop-up box. The disadvantage of using linked files is that printing the whole document may be tedious. Some tools are designed solely for use with computers but others provide the ability to print complete documents with the diagrams. A tool that allows you to add text to the boxes on the flowchart and generate a complete procedure, complete with headings and contents list, is Modeler™ from CASEwise.

## Documentation standards

### Presentation

Documents may be presented in a variety of ways. There are no requirements in ISO 9000 governing presentation standards and therefore the choice is yours! There are several options to consider, many of which only apply to paper versions of a document. In electronic versions, many of these features are provided automatically by the word processor. An example of a documentation standard is given in Annex B. Several documentation issues are addressed in Table 6-1.

## 194 Documentation develoment

| Documentation issue | Suggestion |
|---|---|
| Should each page carry the company logo? | This is not essential but helps to give the document legitimacy if it does. |
| Should each page carry approval signatories? | This practice is unnecessary and a waste of time. Documents need to be approved. How you denote approval is up to you. (See later.) |
| Should each page carry number of pages in document (e.g. Page 1 of 6)? | A good idea since it helps users detect whether they have the complete document. (See later.) |
| Should the pages be printed on coloured paper? | Nice but not essential. The more you spend on presentation the more barriers you put in place for people to resist change. |
| Should printed documents carry the label 'Uncontrolled Document'? | Totally unnecessary. If it is your policy that the electronic version be the only authorized version then marking paper copies is an added burden. If it is done automatically then perhaps there is some merit as a reminder to staff. If it has to be done manually then don't do it. (See later under *Document control*.) |
| Should each document carry a title page? | If the documents are to carry approval signatures and a distribution list then a front page may be necessary for this information. If not then what purpose will it serve? |
| Should reference codes be added to each page? | Yes, unless the documentation is entirely electronic. A lost page even carrying page numbers will remain lost without some identification to its origin. |
| Should each document bear a date? | Yes, without doubt, as it indicates the age of a document. |
| Should the revision status be added to each page? | Yes, without doubt. If you will be replacing pages in a revision then the revision number is vital for distinguishing between old and new versions. |
| Should revision numbers be used in addition to dates? | It is not essential but advisable. A date can indicate revision status, but only if you have the previous version alongside. A number clearly shows that there were previous versions. A date does not indicate this unless two dates are shown: the original release date and the current revision date. (See also under *Document control*.) |
| Should each document carry an amendment record? | Yes, but it must indicate the nature of change and reference the approved Change Notice. |
| Should the document title be on each page? | If you use a title page then this is unnecessary. If you don't use a title page then, obviously, it is necessary. |
| Should the document contain forms that are referenced within? | If they can be printed full size and used, then they could be included. Otherwise, directions to where they are located should be given. If all forms are contained in a separate volume or directory, then one instruction in the Policy Manual will do this. Do not reduce the size of forms and place them in the procedure as this reduces maintainability. If the form changes you have to change two pieces of information not one! |

Table 6-1 *Documentation issues*

The questions in Table 6-1 need to be answered before you commit yourself to document production. Some, of course, may be left until documents are ready for approval but, once a document is approved, and these issues remain undecided, you will give yourself more work, bringing all documents to the same standard. It is not a nonconformity with ISO 9000 but consistency in presentation sends a certain signal to the readers.

> **THE WRONG SIGNALS**
>
> If those who are managing the quality system can't get it right first time then they have no right to expect us to.

### Style

The provisions within documents can be written in different styles, each having a different impact and purpose. Documents should be produced using the third person (it) not first person (I or we) although, if the document is a proposal, an organization may write it in the first person plural. Style is important as it influences how a reader will respond. If you write in an inappropriate style for the particular document, you may not cause to happen what you had intended. If you have found the best way of carrying out a task you would want the task to be carried out that way and not allow the documentation to offer too much flexibility. The style you use will dictate the effectiveness of the documentation. Remember that policies and procedures need to cause things to happen. They are not history books – statements of what we once did. They should signal your commitment. The documents do have some legal significance – they disclose your policies and practices, not your dreams and ambitions. If you need to defend your actions either to an auditor or a court of inquiry in a product liability dispute, incident or accident, the language you use in your documents can make a lot of difference to how your actions and decisions are judged.

- Were the instructions clear enough for someone to follow?

- Could they have been misinterpreted?

- Were they ambiguous?

- Did they imply options where no options were permitted?

- Were they sufficient to enable performance to be judged?

Statements can be written using the auxiliary verbs 'shall', 'must', 'should', 'will' or 'may', or be written using no auxiliary verbs as instructions or descriptions.

### The mandatory style 'shall'
Statements using the word 'shall' indicate a provision that is binding, such as: 'Design reviews *shall be held* at defined stages during product development.' To the organization's staff, this statement implies that there are no options; design reviews have to be held. The staff are under an obligation to conduct design reviews. However, there are no penalties should circumstances dictate reviews to be inappropriate.

An alternative to 'shall' that has the same effect is 'are to be', such as: 'Design reviews *are to be held* at defined stages during product development.'

### The advisory style 'should'
Statements using the word 'should' indicate a provision that is advisory, such as: 'Design reviews *should be held* at defined stages during product development.' To the organization's staff, this statement implies that it is only a recommendation and not an instruction. Design reviews ought to be held but they don't have to be. There is no obligation to hold design reviews. It tends to suggest that they should be held only if there is sufficient time or if there is no alternative method to verify the design. The advisory style is suitable for guides and reference documents, not for procedures. If you intend to provide advice in a procedure or a standard you can make it a rule that a person has to have 'just cause' to choose an alternative solution, thereby leaving the reader to judge which solution of those provided would be appropriate. When you say to someone: 'You should conduct design reviews,' then this is advice which you wish them to follow but the decision is theirs. In such cases you will probably not be their superior, otherwise you would have said: 'You shall conduct design reviews'. If you say to someone, 'You may conduct design reviews,' you are giving them the option to conduct design reviews if they want to, implying there are other alternatives.

### The emphatic style 'must'
Statements using the word 'must' indicate a provision that is compulsory, such as: 'Computer viruses *must not be allowed to* enter the information environment.' To the organization's staff, this statement implies that there will be unfavourable consequences for the individual if a virus is introduced into the environment. 'Must' statements imply that there is no choice but to comply. The same effect can be produced by notices without using the word 'must', such as 'No Smoking' or 'The consumption of food or drink in this area is prohibited'.

### The future intent style 'will'
Statements using the word 'will' indicate a provision of intent and not necessarily an obligation: for example, 'Design reviews *will be held* at defined stages during product development.' To the organization's staff this statement has less impact than using the word 'shall'. Staff are less drawn to action by 'will' statements as they may perceive them as intentions for the future and not the present. There is also no urgency in the 'will' statement; the urgency comes from the way someone says it. An example which has the opposite effect is: 'Design reviews *will be held* with subcontractors when warranted by

design complexity.' When read by the organization's staff, this statement has more impact and places staff under an obligation to conduct design reviews when judged that the conditions are appropriate.

### The optional style 'may'
Statements using the word 'may' indicate that a provision is optional, such as: 'Design reviews *may be held* with subcontractors when warranted by design complexity.' To the organization's staff, this statement has less impact and permits them to use their own judgement to conduct design reviews regardless of design complexity. Another example is: 'Changes *may be made* by providing a marked up copy of the document or by detailing the changes on the Form.' This statement provides a choice to the reader as to how the changes might be conveyed.

### The instructional style
These are statements that use none of the auxiliary verbs and contain instructions, such as: 'Send out the agenda for the design reviews two weeks before the review meeting date.' This style is suitable for operating procedures where the responsibility is stated in the control procedure. In this way, operating procedures can be a series of instructions or steps, as might be found in a self-assembly kit of furniture. For example:

---

1  Open the box and check the contents against the list enclosed.

2  Fit the dowels (item 1) into side panel (item 2) using glue supplied.

3  Fit item 3 (4 off) into item 2 as shown in Figure A.

4  Join top and bottom panels (item 4) as shown in Figure B and press dowels firmly into place. Excessive force will cause permanent marking of the surface.

---

### The descriptive style
Statements that also use none of the auxiliary verbs are statements of fact, such as: 'Design reviews *are held* at defined stages during product development' and 'All products *are subject* to inspection on receipt.' To the organization's staff, this statement may have the impact of 'So what?' The style is appropriate in an exposition of your company's operations. When used in policy and procedure statements, however, the style does not command action. It does not cause things to be done, unlike the 'shall' and 'are to be' statements.

### Postscript on style
Style is often a matter of choice and dependent on the organization's culture. Some organizations prefer not to be given orders and a 'shall' statement is akin to an order. In others they are more comfortable writing in 'will' statements but unaware of the consequences this has when written into policies and procedures. In summary:

- Use 'shall' statements for policies, control procedures and standards.

- Use 'should' statements for guides.

- Use 'must' statements for compulsory issues.

- Use 'will' statements for proposals.

- Use 'may' statements only for conditional issues.

- Use descriptive statements for expositions.

- Use instructional statements for operating procedures.

**Layout**

*Layout* is concerned with the page disposition, font, spacing, margins, headers and footers, etc. Consistency is the key rather than prescription. Adopt a suitable layout then stick to it. Many companies have a house style with regard to these features. It is sometimes too constraining and unnecessary for internal documents. Some points to consider are given below.

*Margins*
Make the binding margin wide enough to allow for punched holes. Check your hole punches before setting the dimensions. Some people go to the extent of buying in prepunched paper but unless you have a regular use for it, this is not necessary. Make the top and bottom margins wide enough to cater for variation in photocopying. If the documents are to be viewed on screen, then margins are also important (especially if the software does not have a zoom function). Even with a zoom function, it becomes irritating to have to reset it every time a new document is displayed.

*Font*
Don't use fancy fonts! Arial or Times New Roman are readable by anyone. You are not producing a work of art but working documents. Screen sizes can be enlarged by the word processor so set a font size that is readable when printed. Be consistent with the fonts so that reading is easy on the eye.

*Spacing*
Provide line and paragraph spacing that aids reading and minimizes chances that lines will be missed. Separate the headings from the text so that the change in subject matter is clear.

## Headers and footers

If you are not going to have a title page, the header can carry the document identification data such as Title, Reference Number, Revision Status, Date and Page Numbers. Alternatively you can place the title in the header and the other data in the footer.

## Titles

Document titles are labels that you give to communicate the subject matter of the document to its readers. We have difficulty relating numbers to content but can easily relate titles to content. Titles, however, do have limitations: too brief and they mislead, too long and they confuse. The title should convey the subject and its scope in a few choice words. There is no need to include words such as Document, Procedure, Standard, Specification etc. unless it is not obvious. For instance national standards have the code for standard in their numbering system, so making it unnecessary to put the word 'standard' in the title of every standard. If you have two procedures for purchasing, for instance, each should carry a qualifying word in the title, such as 'Product Purchasing' and 'Capital Equipment Purchasing'.

## Dating conventions

The date denoted on a document should mean something. It can be:

- The date of creation

- The date of release

- The date of approval

- The date of print

- The date it becomes effective

Whichever convention is used, it should be defined so that users know what it means. With word processors, you have many choices. If you choose the 'print date' then every time you print the document the date changes whether or not you change the document! The 'last save date' is the best option if you want it to be automatic. You may also find it useful to insert the creation date, thereby giving the reader an indication of the maturity of the document. The only problem here is if you need to recreate the document as a result of changes (in structure, numbering or simply a formatting problem) a new creation date will be inserted if you use the automatic features. This problem is avoided in Lotus Notes with a 'create new version' feature that retains the previous version's formatting.

## Document coding

Document coding systems take a variety of forms. The essential criterion is that the code given to a document is unique to that document so that references to it will lead the seeker to the correct document and not to some other document. The document title is not suitable as the sole identifier despite that fact it may be possible to accommodate a wide range of titles before difficulties arise. There are several options to choose from.

### Aligning document codes with the standard

This is not recommended: firstly, because the standard will change and, secondly, it shows that your motivation is compliance with the standard rather than improving the business. Examples are:

> QP 4.5 for the Document Control Procedure
>
> QP 4.6 for the Purchasing Procedure

This may work for some procedures but not for all. There will be instances where you have a need for a procedure that does not fit an element of the standard. It also does not work for documents other than the Control Procedures.

### Aligning numbers with the Policy Manual or Quality Manual

This is similar to the previous option but has a wider scope, as it is not tied to the standard. However, there will be documents that apply to more than one policy. Again it doesn't always work, except for the Control and Operating Procedures. Derived documents would need to adopt another scheme.

### Sequential numbers based upon subject grouping

With this approach, a range of subject groups are established (either matching functions or processes) and these are coded. Examples are DD001 for Design Department, PD001 for Production Department, QD001 for Quality Department. Alternatively, companies tend to assign the letters QP for the common Quality Procedures, with the Work Instructions following a departmental scheme.

Another variant on the same theme is to use QP-DD-001, indicating the document is a quality procedure that belongs to the Design Department. One problem with adopting functional coding is that should a function change and tasks transfer to another function, the documentation would also have to be renumbered. With process codes this does not occur as processes normally remain the same whichever function carries them out.

### Site identifiers

Where there is more than one site or division, companies add additional codes, so that QP-DD-001 becomes AD-QP-DD-001 for the design procedures of the Analogue Devices Division. The letters can get beyond reason. Some documents carry divisional identification through logos to reduce the length of the document code.

## Document type identifiers

QP in the above example is a document identifier and this concept can be extended to cover a wide range of documents. You can group them solely as document types or insert separators to identify functions. Some examples were given previously in this chapter. Further examples are:

*CP001 for a Control Procedure*

*OP001 for an Operating Procedure*

*BF001 for a blank form*

If you include function indicators then CP001 may become CP-DD-001 indicating that the document is a control procedure belonging to the Design Department. The function code could be used to identify processes so that you are not constrained to the organization. Where documents are common to more than one function the letters CD for 'common document' could be used. In small organizations, functional identity should not be needed.

Alternatively, a numerical sequence can be used in place of letters; for example:

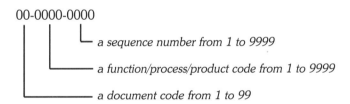

00-0000-0000
- a sequence number from 1 to 9999
- a function/process/product code from 1 to 9999
- a document code from 1 to 99

This method has the advantage of collecting documents with common features under one number. Although rather long, staff become accustomed to the codes and recognize groupings after the system has been in use for some time. Such a system is only necessary in complex organizations with wide product ranges.

## Date identifiers

With on-line documentation you can use the 'creation date' as the unique document number. The only problem with this is that it is meaningless and is unlikely to be remembered. A typical date code might be 970312-082916 meaning 29 minutes and 16 seconds after 8am on 12 March 1997. Further identification could be provided by adding document codes so that the document number becomes CP-970312-082916. This type of number is better hidden in the database as its use as a reference code does not permit listing by number. In a list you cannot account for the quantity of documents, unlike the simple sequence number. It is also clumsy if you want to add reference numbers in a table or index in place of a title. The number is almost as long as the title. If you want to list documents coded in this way in a certain order, then you will need to code the title to use the sorting features of the software.

## Quality characteristics of documents

Any document has a distinct purpose and, as such, must fulfil this purpose. It will not fulfil the intended purpose unless it possesses certain characteristics. Many of the characteristics are not formally stated but are expected by users. Users expect a document to be:

**Consistent**  They expect to find unified communication both within the document and within other documents with which it is related. They do not expect to find inconsistencies. A message in one part of the document should not be in conflict with a message in another part, neither should it conflict with messages in other documents from the same set.

**Maintainable**  They expect documents to be easy and economical to maintain. When a statement is amended it should not be necessary to seek other instances of the same statement for amendment other than for a single word. They also expect that changes in the organization that do not affect practices would leave the related documents unaffected.

**Reliability**  They expect to rely on the information being accurate and legitimate – that messages had been checked for their validity before release. They also expect that the document would not change unless the revision status was also changed. They expect a 'shall' or 'must' statement to be essential and not to find some managers easily giving waivers. Likewise, they expect a 'should' or 'will' statement to signify an advisory or conditional action and not to be reprimanded for taking an alternative action.

**Flexibility**  They expect documents to allow innovation where it is practical and for the documents to allow minor variations in practice where the method is immaterial to the result.

**Vulnerability**  They expect the documentation to omit any requirements that could not be achieved by staff of average ability using available resources. They do not expect to find that only superhuman staff could meet the requirements!

**Compliance**  They expect the documentation to reflect compliance with national and international standards, applicable laws and statutes so that following them would provide immunity to prosecution.

| | |
|---|---|
| **Usability** | They expect the documents to provide the right information for them to determine the right things to do and to enable them to do these things in the right way the first time and every time. |
| **Traceability** | They expect the documents they use to be traceable to governing requirements, policies or objectives that cause each document's existence so that they serve a useful and economical purpose in the organization. |

Documents should be produced with these characteristics in mind and, when submitted for review and approval, such characteristics should be included along with more specific content requirements.

## Immunity to change

If your documentation is paper-based then it may soon become outdated as changes in the business take place. It is frustrating to find that no sooner have the documents been approved and issued, than business changes are made which affect some or even all of the documents in the quality system. This is a big problem with paper-based systems but less of a problem with on-line documentation. Both types of documentation need to be changed but, in the case of electronic documents, changes can be made in some cases by making one change and employing macros to implement the change in all the documents affected. Such changes are an unwanted diversion if they do not materially affect practices. Cosmetic changes such as changes to names, locations, job titles, organization charts etc. may have no effect on the processes. Who does what is, after all, unimportant. What is important, however, is that whoever performs the job they should be qualified to do so. The effectiveness of the process depends upon employing the right materials, people, procedures, environment etc. to convert inputs into outputs. The right people are not people who happen to work at the current time in a particular department but the people who have the right qualifications and expertise to warrant the responsibility and authority vested in the task. If the manager is qualified to perform a task (by virtue of his/her education, training and experience) and a member of staff goes sick, the manager should legitimately be able to perform the task without violation of the documented policies and practices. This won't be possible if you have included the names of staff or used job titles in your documentation. A person may be capable of performing many roles and, providing he/she is suitably qualified, the documentation should not prevent legitimate action or decision.

> **KEY MESSAGE**
>
> The department in which people work is immaterial to the effectiveness of the process.

If you reorganize the building and move offices and workshops this should not affect the documentation. If you move to a new site or expand onto an additional site the documentation should not be affected if the processes remain the same.

To create immunity to change in your documentation you should:

- Use roles in place of job titles, department titles and names.

- Use generic place names for locations that can be carried with the process wherever it is carried out.

- Don't put room numbers, geographic locations etc. in documents. If you need to add such data then place it in a reference document so that only one document has to be changed.

- Limit dates in documents (other than document release dates) to those that define objectives, plans or schedules.

Examples are given in the table below.

| Function | Alternatives |
| --- | --- |
| Maintenance Department | Maintainer<br>Maintenance Authority |
| Design Department | Designer<br>Design Authority |
| Purchasing Department | Purchaser<br>Purchasing Authority |
| Quality Department | Auditor<br>Investigator Release Authority |
| Quality Manager | Quality System Manager |
| Software Development Group | Programmer<br>Systems Analyst |
| Goods Inwards Inspection | Inspector<br>Tester |

## Creating the quality policy

The purpose of the quality policy is to define the organization's overall intentions with respect to quality in such a way that it will guide the actions and decisions of all those within the organization.

The policy constrains actions and decisions by channelling thought processes along a path that will lead consistently towards the achievement of quality. It should cause managers to take decisions that enable their objectives to be achieved and at the same time achieve product and service quality. The quality policy is not a procedure. It should not define how quality will be achieved for specific products and services but define an approach that everyone will pursue which will ensure only quality products and services are provided by the organization. Clearly, such a statement needs to be much more than a slogan. 'We will do it right first time', 'We will be the best' and other vague statements will not achieve this goal because these slogans raise questions such as 'What is *it* ?', 'What is *right* ?', 'The best at *what* ?' and 'What is the *best* ?'

Some examples of the subject matter for quality policies are given below.

---

### SUBJECT MATTER OF THE QUALITY POLICY

The policy statement should define the organization's intentions with respect to:

**Satisfying customers** — Do we intend to convert customers to accept what we can deliver or establish what they require and aim to exceed their expectations?

**Treatment of customers** — Do we intend to put our needs before those of our customers or place customer needs above our own?

**Treatment of suppliers** — Do we intend to form adversarial relationships with our suppliers or form partnerships?

**Treatment of employees** — Do we intend to exploit our employees or treat them as our most valuable resource?

**Quality control** — Do we intend to control by inspection or by prevention?

**Quality improvement** — Do we intend to remain at our present level of performance or seek year-on-year improvement?

**Staff development** — Do we intend to rely on self development in our staff or develop and motivate our staff so that we gain exceptional performance from average ability?

**System certification** — Do we intend to use certification as a marketing ploy or as a means for raising standards of performance?

**System use** — Do we intend to use the quality system only to retain certification or to enable us to run an efficient and effective business?

---

Addressing these topics should take no more than one page but if it does then so what? It doesn't matter. Many companies believe that the policy has to be learnt by all the staff so that they can repeat it to the auditors when they ask. This is not so. The standard does not state that everyone shall quote the quality policy. It says that *staff at all levels shall understand the quality policy*. This is quite different. To understand the quality policy you need to act in a manner that shows you understand the policy. If you say you will meet customer requirements on time every time then why did you fail to make the last five deliveries on time? Why did you ship nonconforming product last month? Why have you so many customer complaints? Why do you allow your staff to deviate from the approved procedures?

If every company honoured its stated quality policy there would be no customer complaints, no late shipments, no defective products. So most companies are not doing what they say will do! If your quality policy aims for perfection, which many of them do, one failure, however small, is a failure to meet the quality policy. Be realistic! Provide for improvement, provide for the odd failure but promise you will strive for continual improvement in performance. In that way you give yourself some credibility[1].

## Creating the Policy Manual

### Philosophy

ISO 9001 requires a Quality Manual and in ISO 8402, a Quality Manual is supposed to contain the quality policy and describe the quality system. As Alan discovered in Chapter 1, many quality manuals do not do this. They merely respond to the requirements of ISO 9000. ISO 8402 defines a quality system as *the organization structure, processes, procedures and resources needed to implement quality management* Therefore, to describe the quality system you would need to describe the organization structure, processes, procedures and resources needed to implement quality management. These aspects were addressed in Chapter 2. A description of the organization structure should be more than a chart and should cover the vision, values, mission and policies that shape the organization to fulfil its mission. The processes should be the business processes drawn from the system model. The procedures should be included if a few lines or referenced if in separate documents. If a coherent strategy has been adopted, reference to the control procedures will enable all others to be located (see below). The resources should be described in terms of the facilities, equipment and personnel utilized to execute the business. The descriptions do not have to be in the manual provided they are contained in reference documents.

### Purpose

The purpose of the Quality Manual is to contain descriptions of the policies and practices that enable the organization to meet its customer requirements and business objectives.

---

[1] See *ISO 9000 Quality Systems Handbook*, Part 2 Chapter 1 for further examples.

As the descriptions of the practices may be extensive if they are to describe how activities are to be carried out, it is often more practical to separate policies from practices. ISO 10013 gives an example of a Quality Manual that includes both policies and practices but the example provided is limited to Internal Quality Audit (a relatively simple and uniform process). When one starts to describe the core processes of the business in terms of the policies and practices that are employed, there may be many practices extending over several pages. Many organizations therefore produce one manual that contains the policies and one or more that contain the procedures, standards, guides, forms etc. The manual that is limited to the policies is, therefore, in principle, a Policy Manual and so the label 'Policy Manual' is more appropriate than the label, 'Quality Manual'. It would not be a manual if all that it did was to respond to the requirements of ISO 9000. However, by referring to all the implementing documents, the policy document becomes more than a statement of policy, a user manual for the quality system.

## Outline

If we adopt the philosophy outlined previously we can come up with a manual that means something to the organization, not just a document produced for the benefit of the auditors. If we model our manual around the business then the foundation for the manual would be the System Model that we developed earlier (see Chapter 5). By taking each business process and by developing policies that will govern the process, we can develop a manual more aligned to the business. Whilst every manual will differ (as every organization is different), a generic model can be created which can be tailored to suit any organization (as illustrated in Table 6-2).

The section on Business Management is expanded since the work processes are fairly common to any business, although in practice the scope will depend on the model developed. Resource management, again, is a common business process. At the policy level the subdivisions may well be as shown. It is at the procedure level that the processes will differ.

## Content

Within the sections of the manual, except for the first two sections which introduce the manual and the business, the content should address policies and practices. It should describe the operating policies that set the boundary conditions for actions and decisions in each process and then reference the practices that implement these policies. If the practices are simple, then obviously you can include them rather than reference them. When formulating policies we need some criteria to help us determine whether a statement is or is not a policy.

| Section | Scope |
|---|---|
| 1. Introduction | • Purpose of the manual and the system<br>• Scope of the manual and the system<br>• Applicability of the system<br>• Definitions |
| 2. Business description | • Nature of the business<br>• Customers and suppliers<br>• System Model |
| 3. Business management | • General overview<br>• Strategic planning, vision, values, mission<br>• Business planning, objectives, targets<br>• Organization development, roles, responsibilities, structure<br>• Performance Review |
| 4. Resource management | • General overview<br>• Resource Planning<br>• Resource Acquisition<br>• Resource Deployment<br>• Resource Maintenance<br>• Resource Disposal |
| 5. Quality system management | • General overview<br>• Quality system development<br>• Quality system implementation<br>• Quality system maintenance<br>• Quality system evaluation<br>• Quality system improvement |
| 6. Business process A | • General overview<br>• Work process 1<br>• Work process 2 etc. |
| 7. Business process B | Ditto |
| 8. Business process C etc. | Ditto |
| 9. Matrices | • Function matrix<br>• Location matrix<br>• Compliance matrix |

Table 6-2 *Policy Manual structure*

## Formulating operational policies

Any statement made by management at any level which is designed to constrain the actions and decisions of those it affects is a *policy*. Managers issue policies on topics where there is a choice of action or decision (that is, choice between right and wrong). There will be a right way of doing something in your organization and a wrong way, which may not necessarily be the same in other organizations.

Policies serve to guide the actions and decisions required to achieve objectives and are not therefore objectives in themselves. Policies set boundary conditions so that actions and decisions are channelled along a particular path in pursuit of an objective. Many see policies as laws to be met. They are laws, in so far as an enabling mechanism. Policies enable management to operate without constant intervention and, once established, enable others to work within a framework without seeking decisions or guidance from above.

Staff do not work to policies but, in fact, work in accordance with procedures which themselves direct actions and decisions within the framework of the stated policies. In order to make the decisions required of the procedures, staff will often need to know the company policy on a particular subject, such as procurement, recruitment, release of product, licensing agreements and agreeing design changes. Staff will also need to know if they can or cannot do something and, if so, what criteria should they satisfy?

When one deviates from procedure one may not in fact be violating a policy, as the procedure may describe one of several ways of doing something. For instance, a policy can remain unchanged when processes and procedures change from being manual to automated. Although people no longer do the tasks and make the decisions, nonetheless these actions are still taken.

There are many sound reasons for documenting your operational policies:

- To translate corporate policy into practical terms which can be implemented through procedures.

- Every job has constraints surrounding it – without written policies people would be left to discover them by trial and error, the organization would become a disorganized mess, its managers lacking any means to direct and harmonize their staff's activities.

- Policies enable managers and their subordinates to be left in no doubt about what they are actually responsible for, the boundaries within which they have to work and the demands upon them to which they are expected to respond.

- Policies set clear boundaries for people's jobs so that everyone knows in advance what response they will get from others when making decisions.

- Policies create a baseline to which subsequent change can be referred and ensure changes in the way things are done are clearly defined.

- Policies enable managers to determine whether a subordinate's action or decision was simply poor judgement or an infringement of the rules. If no rule existed, subordinates cannot be criticized for using their judgement, however poorly it is used. If a rule exists, one has to establish whether it was accidentally or deliberately broken, for the latter is a disciplinary offence. Without written policy no one knows where they stand and any decision may create an unwanted precedent.

- Policies provide freedom to individuals in the execution of their duties to make decisions within defined boundaries and avoid over-control by managers. If people are uncertain about where the limits of their job lie they cannot feel free to act. Without a clearly defined area of freedom there is no real freedom.

- Policies enable management to exercise control by exceptions rather than over every action and decision of their subordinates and therefore enable self-control by subordinates.

- Policies enable managers to control events in advance. People know the rules before acting and so are more likely to produce the right result first time. Without policies, one is forced to control events after something has happened to cause dissatisfaction. Alternatively, one would have to be on the scene of the event to respond as soon as the situation approaches the limits. This is a costly use of managers' time.

However, one does not need to write everything down, as policies are needed only for important matters – that is, where the question of right or wrong does not depend upon circumstances at the time, or when circumstances only rarely come into the picture.

Policy statements should pass the following tests:

- They should define what can, or cannot, be done.
- They should not say how something should be done.
- They should provide overall direction so that staff know the intentions of management.
- They should provide rules that constrain decisions.
- They should allow managers to intervene by exceptions, not as a routine.
- They should clarify either directly or indirectly the organization's intentions as to compliance with the requirements of ISO 9000.

## Operating policies

Whilst it is not possible to provide a standard Policy Manual that will be suitable for all companies, examples illustrating parts of a Policy Manual are provided in Chapter 9. These address the principle requirements of the standard in a form that matches the outline described above.

## Compliance matrix

If you respond to ISO 9001 using a compliance approach, there is no need to provide a cross-reference table showing the relationships between the sections of the Policy Manual and the standard. If you adopt the approach described in this book, then such a matrix will provide a useful guide not only to auditors but also to any reader wishing to establish such relationships. There are several ways you can do this:

- Cross reference by element to sections of the manual.

- Cross reference by clause to sections of the manual.

- Cross reference by document requirement to sections of the manual. (See Annex A of Chapter 5.)

The last option is more comprehensive and hence introduces maintenance problems. The first option is too coarse to be of any value so the middle road is the preferred course to take.

## Functional matrix

Whichever approach you take for the Policy Manual – whether you have chosen an element-based approach or a business-process approach – the relationships between the policies and practices and the allocation of work within the organization may need further clarification. The functional matrix should show the processes in which each function has responsibility. An example is provided in Table 5-1 of Chapter 5.

## Location matrix

If your organization is dispersed over several sites, a location matrix may be necessary. A location matrix is particularly useful to the auditors when planning their audits. The location matrix should show the sites on which each function has a presence. If their presence varies then a coding convention can be used to depict the type of personnel present.

For example:

- **S** Sales
- **M** Management
- **O** Operational
- **S** Support

If some sites have few people present then it may be necessary to display the number of people on each site. The more detail you provide, the more maintenance is required.

## Creating procedures

**What are procedures?**

A procedure is a sequence of steps needed to execute a routine task. ISO 8402 defines a procedure as *a specified way to perform an activity*. It prescribes how one should proceed in certain circumstances in order to produce a desired result. Sometimes the word 'procedure' can imply formality and a document of several pages but this is not necessarily so. A procedure can be five lines where each line represents a step needed to execute a task.

Quality system procedures are a particular type of procedure. They implement the operational policies and regulate processes which produce an output (the quality of which is essential to the business). People, rather than procedures, achieve quality. People, rather than procedures, make decisions. So you could have the best documented procedures in the world and still not achieve quality. There has to be a combination of people and documented procedures for you to achieve the desired quality.

The standard only refers to procedures as the category of quality system documentation. If we use the term 'documented practices' we have a wider choice as to the types of documents we put into the quality system. Many documents are not procedures. They do not tell us how to proceed or specify a way to perform an activity. They specify criteria we must meet or provide guidance in carrying out a task. They may also give examples or define rules to follow.

## Types of procedures

There are various types of procedures:

- **Divisional procedures**, which apply to more than one division of a company and regulate common activities.

- **Control procedures**, which control work on product or information as it passes between departments or processes. These should contain the forms which convey information from department to department and reference the operating procedures which apply to each task.

- **Operating procedures**, which prescribe how specific tasks are to be performed. Sub-categories of these procedures may include: test procedures, inspection procedures, installation procedures etc. These should reference the standards and guides (see below) which are needed to carry out the task, document the results and specify the forms to be used to record information.

The relationship between these procedures and the policies is illustrated in Figure 6.1.

## Procedures and instructions?

There is often confusion between the terms *procedures* and *instructions*. Now that the term *work instruction* has been removed from ISO 9001 the confusion should be reduced but there are many companies that still use this distinction.

Differentiating between procedures and instructions implies that there is a difference. Procedures are practices, work instructions are not. They merely contain instructions to do work, and although many organizations produce documents they call work instructions they are in fact procedures because they specify a particular way to perform an activity. An instruction conveys knowledge, information or directions for the purpose of performing work. Procedures also convey knowledge, information and directions but they relate to how certain tasks are to be performed and specify who is to perform them. For example, you may issue an instruction for certain goods to be packed in a certain way on a particular date and for the package to be marked with the contents and the address to which it is to be delivered. So that the task is carried out properly, you may also specify the methods of packing in a procedure. The procedure would not contain specific details of the particular package – this is the purpose of the instruction. Procedures by themselves do not cause work to be carried out. This is the role of instructions. For instance, a purchase order is an instruction to a supplier and the method of producing one would be defined in a purchasing procedure. Similarly, Change Requests, Amendment Instructions, Engineering Orders and Print Requisitions are all instructions that cause people to do work and hence are work instructions rather than procedures.

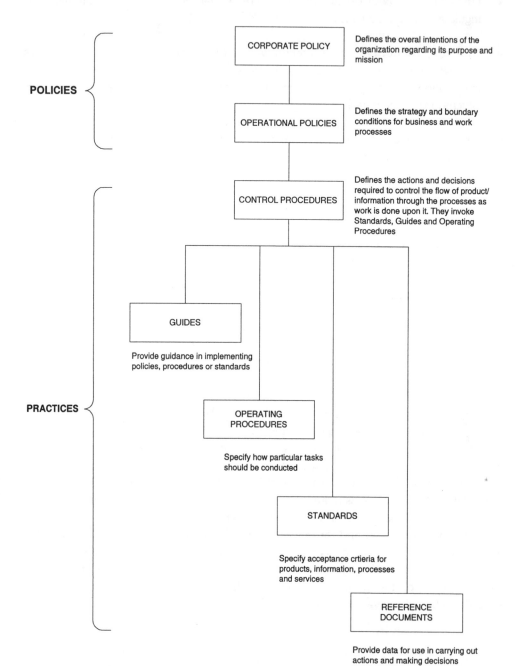

Figure 6.1 *Relationships between policies and practices*

Work Instructions cause work to commence or to stop, procedures define how such work should be performed.

Another use of the term 'instructions' is in the performance of activities relating to a specific task, such as Assembly Instructions, Maintenance Instructions, Operating Instructions and Cleaning Instructions. These prescribe the activities to be performed by a person when carrying out a specific task and, as such, define how the task is to be carried out. They do not define when, who or how the activities are to be performed. For example, an assembly instruction may state that panel A is to be fixed to panel B using the screws provided. The instruction will not tell you how to use a screwdriver but it may tell you to be careful not to damage the outer surface or not to overtighten.

## Size of procedures

If you have a small business and only one way of carrying out work then your system will tend to be small. Your quality system may be described in one document of no more than 30 pages. On the other hand a larger business may require several volumes and dozens of procedures of over ten pages each to adequately describe the system. Control procedures need to be user-friendly and so should be limited in size. Remember, you can use other documents such as guides, standards and operating procedures to extend what you have written in the control procedures. However, the procedures should not be so short as to be worthless as a means of controlling activities. They need to provide an adequate degree of direction so that the results of using them are predictable. If you neglect to adequately define what needs to be done and how to do it then don't be surprised if staff don't know what to do or constantly make mistakes. It is also important to resist the desire to produce manuals that are impressive rather than practical. Printing the documents on expensive paper with a coloured logo does not improve their effectiveness and, if they are not written simply and understood by a person of average intelligence, they will not be used.

## Guide to procedure development

When deciding what should go into procedures one should concentrate on defining the activities and decisions to be carried out and only prescribe how actions and decisions should be carried out where:

- The method is critical to the result.

- Too much choice can be bad for efficiency.

- It would be unreasonable, unsafe or unreliable to expect staff to commit the knowledge required to memory.

## Structure

There are several models to use for the structure of procedures but many follow a simple structure similar to the following:

- Purpose
- Scope and Applicability
- Definitions
- Procedure
- Reference Documents

There are variations on this theme, with Objectives in place of Purpose, References before Procedure, and a section on Responsibilities. However, it should not be necessary to define responsibilities separately from the procedure text since the 'who does what' should be stated in the text. By including a separate Responsibilities statement, you increase the probability of duplication and conflict. You can identify the functions to which the procedure is applicable in the Scope and Applicability statement.

Control and Operating procedures could have the same structure but Standards and Guides may vary. To use the same structure for all documents you would need to change the section *Procedure* to a less definitive term such as *Provisions*.

A brief explanation of the content of each of these sections follow.

## Purpose

The Purpose statement should address the following:

- What the task aims to achieve
- The reason for controlling the task

It should be in the form:

> *The purpose of the (subject) is to ...*

For example:

> *The purpose of controlling changes to documents is to cause desirable changes to be implemented speedily and prevent undesirable changes from being processed.*
>
> *This document implements the document change policies of the Policy Manual, section 5.4 and responds to clause 4.5.3 of ISO 9001.*

It should be possible to verify that the purpose has been accomplished so that there is some basis to judge the effectiveness of the procedure. If you can't verify conformance the statement will need revising.

Although not essential, the reference to the Policy Manual and ISO 9001 does provide the link to trace documents forward through the system. Likewise, if the document is an operating procedure you could refer to the control procedure in which it is invoked.

## Scope and Applicability

The Scope statement should define what the procedure covers, its range and depth.

The Applicability statement should define what, who and where the procedure applies and what the procedure is limited to. For example:

> *This procedure covers the initiation, evaluation, approval and implementation of all changes to internal quality system documentation except project documents.*
>
> *It applies to all staff using management system documentation on all sites.*

## Definitions

This section should contain a statement similar to the following:

> *All general terms and acronyms are defined in the Quality System Glossary. Those unique to this document are defined as follows:*

In defining the terms keep the statements short and concise. Do not include requirements or methods in the definition.

## Procedure

You could provide a brief description of the process from input to output as a summary of the procedure including any optional paths. This is useful for browsers and helpful for identifying common controls or aspects of the flow diagram that are more easily explained in words.

Include the flowchart at this point as follows:

> *The ... process illustrated in Figure 4.1 indicates the tasks, responsibilities and interfaces associated with its execution.*

The following sections would contain some or all of the following elements:

- Paragraphs describing the actions and decisions required and indicating the role responsible by matching the flowchart in the sequence in which they occur.

- The minimum information and equipment needed to perform each activity or make each decision.

- The criteria for decisions as a list of aspects to be considered or a statement of requirement that the decision should satisfy.

- The criteria for choosing optional routes and the sequence of steps to be taken.

- The entry conditions for starting the process in terms of the minimum inputs and approvals to be satisfied before the procedure may commence.

- The exit conditions for ending the process or task in terms of the minimum outputs and approvals to be satisfied for successful completion of the process.

- The source of information or product needed in terms of from what process, what procedure, what person (role) or organization it comes.

- The routing instructions for information or product emerging from the procedure.

- Any precautions needed to prevent incident, accident, error, problems etc.

- Any recording requirements to provide evidence of actions, or decisions, or to enable traceability in the event of subsequent problems.

- Any rules that have to be followed in order to ensure that the task is carried out in a uniform manner and satisfies statutory obligations.

- Controls needed to verify the quality of any products, with feedback loops.

- Controls needed to verify that the process or task achieves its purpose and to verify that critical activities and decisions occur when required.

- Any forms to be completed together with form-filling instructions and responsibilities, the numbering system to be used, and the registers to be maintained.

- Cross reference to other documents in which essential supplementary information can be found.

## Writing the text

When creating the body of the document bear in mind its purpose and the type of document it is.

- Do not go into great detail in a Control Procedure.

- Do not give methods in a Policy.

- Do not give procedures in a Standard.

- Do not define requirements in a Guide.

- Do not specify policies or procedures in a reference document.

The first section of the procedure should address the first box on the associated flowchart. The words in the box should be transposed into a suitable heading. The wording may not be identical but, if it is, it helps create consistency. The second section should represent the second box on the chart, and so on. Where there are decision diamonds, the text associated with the decision might be combined with the preceding activity or as a separate section depending on the significance of the decision.

In writing the text, you are describing what someone will do to execute the activity and, therefore, this needs to follow logically from input to output. For instance, if the task commences on receipt of a product or information then the text should start:

> *On receipt of ... from ... (the task) shall be carried out using ...*

This is the **When** of the procedure.

Then you ask: 'What happens next?' For instance:

- **How** is the activity carried out? Do I do A, B and C or do I follow another procedure? If so, refer to the other procedure in the form: 'X shall be carried in accordance with Y'. If the task is not a sequential task but one which requires judgement, then a procedure would not be necessary. However, a guide that contains various options or routines may be quoted in the form: 'X shall be carried out using Y as a guide'.

- **Who** carries out the task? (Not the name of the person but the role a person performs: the operator, project manager, auditor, maintainer, administrator etc.)

- **What** do I do with it? (Do I file it, store it, move it, dispatch it or what?)

- **Where** do I send it?

Then ask: 'What happens next?' If the task is a decision, you might state:

> *The Y shall be submitted to A for approval and on return it shall be routed to ...*

For approval or acceptance decisions, the text might be of the form:

*On receipt of Y it shall be examined for compliance with the requirements of STXX by A and the results recorded on form BFXX and routed to ...*

Then you should consider **What** happens if? In the event of rejection, what steps are taken. If the (product/service/document) does not meet the specified criteria, it is (action) and routed to (actionee). If a product is rejected, a form may carry details of the reasons for rejection. The form should be identified and instructions provided on its completion if not self evident. Forms are instructions themselves. A box with a title *Product ID* is an instruction to write the product identity in that box.

## References

This section should be used to identify all the documents referred to in the body of the document with the following sentence:

*The table below defines the documents used and generated in the execution of this procedure together with the location, retention period and responsibility for their maintenance. The retention period is indefinite unless otherwise stated.*

| Reference | Title | Location | Retention | Responsibility |
|---|---|---|---|---|
| Reference number of document or document code | Title of document as displayed on it | For electronic documents, the server, database or mainframe<br><br>For paper documents, the library, manual or volume | Months/years to be retained | The role responsible for its maintenance |

ISO 9001 does not require you to define retention times for documents other than records. Opinions seem to differ, however, on what a record is and it would seem logical to stipulate retention times for all documents, as your system would become ineffective if any documents were not available when needed.

If you have referred to those documents used in connection with the procedure by their reference code, it will be necessary to include their full identity and location in the above table. The retention periods for policies and practices should be indefinite and, therefore, need not be stated.

The table above provides more than mere references and therefore a more apt title would be *Document and Data Management.* Placed at the back of the document, the table collects together all references in the text of the document. It should, therefore, only identify documents referenced in the text. If there is a need for a bibliography, this should be a separate section that merely lists the documents concerned. By including all the above data in one table you can refer to the documents in the text by reference number, unless the number is meaningless and too long. In such a case the reference to a title and a document code (CP, OP, ER etc.) may be preferable.

## Organizing the preparation of procedures

The preparation of procedures takes time and careful planning to get right. A common approach is to assign the writing to one person but this has its drawbacks:

- No person will understand everything about the organization.

- There will be a mismatch between procedure and practice.

- Staff will not own the system as they didn't contribute towards it.

- The system may meet the requirements in the standard but may not be workable.

A better method is to nominate several key staff in the organization to join a quality system development team and arrange for them to produce the documentation with guidance from an expert.

Take care to issue a Document Requirement for each document to be developed. Commencing development without a requirement often leads to several iterations and inconsistencies.

## Document development procedure

With this approach you will need one or more procedures that address a number of processes (as defined in Table 6-3). If it is difficult to persuade managers to follow such formal practices, turn the benefits into questions and judge their response wisely before insisting on formal methods.

## Interfaces

It is important that all procedures interface properly with one another. The outputs of one procedure should lead into the inputs of others. Control of these interfaces can be maintained by using the procedure application routines mentioned above. You should not allow uncontrolled generation of procedures. Each one should be carefully planned in the same way as you would plan the design of a product. Every procedure must have a purpose and a place in the system so that the system remains coherent.

| Process | Purpose | Benefits |
|---|---|---|
| Document application process | For authorizing the preparation of new documents, nominating authors and owners | Causes the right documents to be developed and prevents documents being produced for which there is no requirement |
| | | Ensures system development remains under control |
| Documentation standard | For specifying the format, numbering, layout, typing, issue notation, dating and content conventions to which you wish to adhere | Causes consistency and uniformity and prevents conflict |
| | | Provides basis for document approval |
| Preparation process | For describing the routines that authors and typists should follow to produce drafts, forms, tables, charts and flow diagrams | Causes consistency, uniformity and accountability and prevents unnecessary recycling |
| Review and approval process | For detailing the method for obtaining comment on the drafts and gaining approval of those responsible for its implementation | Requirement of ISO 9001 |
| | | Prevents release of unsuitable documents |
| Publication process | For covering the copying and distribution of quality system documents | Ensures consistency, uniformity and accountability |
| | | Causes documents to be issued to the locations where they are needed and prevents acts of omission |
| Implementation process | For bringing new and changed procedures to the attention of staff and for proving the new methods before full scale application | Ensures consistency, uniformity and accountability |
| | | Causes effective implementation and prevents delays |
| Change process | For proposing, reviewing, approving and issuing changes to quality system documents | Requirement of ISO 9001 |
| | | Ensures consistency, uniformity and accountability |
| | | Prevents undesirable change and causes desirable change |
| Filing and storage provisions | For controlling the master material, computer disks, paper copies, current and obsolete issues | Ensures consistency, uniformity and accountability |
| | | Causes removal of obsolete data and prevents loss of current data |

Table 6-3 *Document development process elements*

## Creating the organizational documentation

The quality system has been defined as the organization, processes and resources to implement the quality policy and achieve the quality objectives. It follows therefore that a key ingredient is the organization and its relationship with the processes and resources. Organization development follows strategy[2]. A diagrammatic representation of Drucker's strategic planning and organization development process is illustrated in Figure 6.2. This may help in the development of the Business Management Process referred to in Chapter 5.

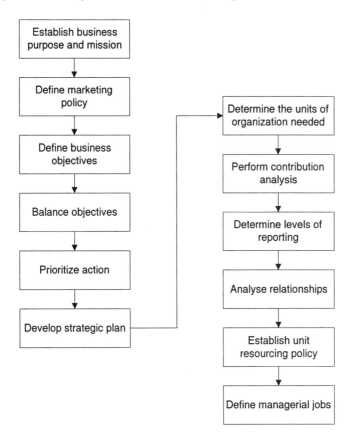

Figure 6.2 *Organization development process*

Several documents are needed to define the organization:

- Charts showing interrelationships between various components of the organization.

- Staff lists showing the names of personnel assigned to a function, with their job titles if appropriate.

---

[2] For a full explanation of the process see *Management* by Peter F. Drucker (1977).

- Function descriptions defining what each function does and the positions appointed to execute the work.

- Job descriptions defining the boundaries of specific appointments in the structure.

- Role descriptions defining responsibilities of the people carrying out roles in the system.

**Organization charts**

Organization charts are needed to show relationships between the components of the organization although many show the reporting channels of the people in the organization. In a free-form organization a functional chart would be better than a people chart. Even where there is no hierarchy, there are still parcels of work to be organized and assigned. In project work, a Work Breakdown Structure (WBS) technique is used to depict how work is divided. An organization chart can then show who has been assigned to perform the work.

An organization chart should not show the names of people as these change more often than the organization. However, many managers take the opportunity to reorganize when staff move on so the chart is never stable. The purpose of these charts should always be defined, as the layout can send inappropriate signals to staff who think it is a hierarchical chart showing levels of responsibility. The functional grouping, however, does give out signals that indicate the relative importance of functions. If the QA function is beneath Production when there are design functions on the chart, it sends out the signal that QA has no responsibility in design. Putting security high in the organization sends out the signal that it is important to the survival of the business. Placing it low, indicates it is of minor importance.

The primary purpose of the chart is to show the relationship between functional groups and is needed when planning change and planning audits. Don't put the charts in the Policy Manual since they are not policies but reference documents. However, the organization chart should be a controlled document since it responds to the requirement in the standard for relationships to be defined.

**Function descriptions**

Function descriptions describe the purpose of a function and its primary accountabilities. The function manager is responsible for managing the function so that it delivers according to its functional description. The objectives would be defined in the annual plan indicating the functions responsible for their achievement. Function descriptions can be placed in the Policy Manual but if there are many functions, it is often more appropriate to locate them elsewhere in the quality system documentation.

These descriptions are high-level documents from which the job descriptions emanate. They are useful to managers, new staff and auditors and should therefore be maintained. There are few tasks for which a function description may be needed but if used for planning purposes or training new staff, out-of-date documents will cause abortive work.

## Staff lists

You don't need staff lists to meet ISO 9000. However, you need to list the names of people somewhere so that managers know who reports to whom and what human resources they have at their disposal. Staff lists can be used in lieu of charts if a list is a more practical way to show relationships.

For the system to function, people need to be able to communicate with one another. The staff list, if made available to all, provides this vehicle. In a small company of up to 20 people they may not be necessary. Whether staff lists should be controlled documents depends on their usage. If a help desk operator needs access to telephone numbers to contact staff to resolve problems, then having out-of-date documents will send the message to the wrong person, thus increasing response time. If your quality objective is to achieve a response time of five seconds and a reply within five minutes, then having accurate staff lists is essential.

## Job descriptions

There are many who believe that ISO 9000 requires job descriptions. What ISO 9001 does require is that the responsibilities and authority of personnel be defined and documented. You can do this through job descriptions, role descriptions or procedures. The disadvantage with job descriptions is that they are often used for purposes for which they were not designed. Most job descriptions are designed for job evaluation purposes to determine a grade and salary and subsequently to measure performance in staff appraisals. They do not, and should not, define all the tasks that an individual will perform. The tasks should be defined in the procedures. The job description should define the results that are to be achieved against which performance in the job can be measured. Hence job descriptions reflect hierarchical positions. You do not need job descriptions if you assign roles.

If you provide job descriptions in response to ISO 9001, they need to be controlled documents because they respond to the requirements in the standard. If out of date, they can create conflict and cause delays when used by inflexible staff. In plants where there are trade unions, obsolete job descriptions can cause demarcation disputes resulting in delayed shipments.

## Role descriptions

The advantage of the role description is that a person can be given any number of roles to perform providing they are qualified to perform them. A role is a group of tasks with a defined objective. A Project Manager is a role since it can be performed by anyone in the organization with the requisite skills and knowledge. The position within the organization this person holds is not relevant. The CEO could be a Project Manager for a very important project requiring significant resources and a fourth-level engineer could be a Project Manager for a far less important project with few resources (see the previous section under *Immunity to change*). If the procedures identify roles rather than jobs or functions, then role descriptions are needed to define them.

Role descriptions should be controlled documents, as the person assigned to the role may not have carried it out before and will need up-to-date guidance to ensure its successful execution. A sample role description is provided in Annex C.

# Creating and using forms

Forms are very much part of the control procedures; in fact, together with a flow diagram, they may *be* the control procedure. You will need to create a mechanism for the control of forms, their numbering system, application and related procedures. Every form should be related to a particular procedure. A form without a procedure to cause its use is not part of the quality system.

## Paper forms

It is the content of the blank form that you should control rather than the piece of paper. If you have a large stock of existing forms it would seem uneconomic to destroy these simply because they are not traceable to the quality system. You have a choice:

- Scrap existing forms and produce new forms on a computer using a word-processing or database package and network the computers.

- Scrap the existing forms and create new paper forms with form number and revision status.

- Include the existing form in the procedure and apply a form number and revision status on each blank using an ink stamp.

- Provide form content requirements in the procedure and use the specific form name and deregulate the forms.

In the case of the last option, control is not as robust but one can check forms against procedures at any time to verify conformance. It also allows forms to be laid out to suit the amount of information provided. There are some drawbacks, however. Where forms pass between departments, allowing individuals to vary the layout may cause processing delays and staff irritation. The layout of forms which pass between divisions and companies should remain uniform to avoid unnecessary disruption. Too much flexibility will almost certainly increase the likelihood of error.

**Electronic forms**

The advantage of electronic forms is their security and to some extent their flexibility. The electronic forms can include all the form-filling instructions and so remove the need for a separate guide or procedure. Provisions can also be made for storage of completed forms in a database so that their location is secured. Such forms provide for data to be amended (unlike paper forms which either have to be regenerated or show where data has been struck out). With a good forms package, password protection can be provided for each text-entry block. These controls, whilst giving good security, can reduce flexibility as they cause only the designated personnel to provide the entry data. There may, therefore, be a temptation for staff to share their passwords so that the process is not interrupted when they are absent.

With paper forms, approval is often provided by signatures and this could be applied in electronic form but is unnecessary as password identification control can be employed. However, security does depend upon users not disclosing their password to others. If a person forges a signature then this is fraud and should be illegal in your quality system. However, the disclosure of personal passwords is not fraud but irresponsibility and the fault of the owner not the forger unless the password was illegally obtained. Electronic forms are therefore more difficult to control.

**Content of records**

There are various types of records but, in addition to the specific aspect to be recorded, records should make provision for the following, as appropriate:

- Record title
- Name of originator
- Creation date
- Location of origin
- Reference to any source documents

- Name of approval authority
- Approval signatures
- Date of approval
- Routing instructions

## Summary

In this chapter we have addressed the development of documentation following the identification of documentation needs in Chapter 5. Several key messages were identified.

- The aim of documentation is to communicate so that users can rely on information being current, accurate and relevant to their job.

- The quality system is a documented system not a system of documentation.

- Quality system documentation should be derived from the system modelling process such that all documents are traceable to a requirement which provides a reason for their existence.

- Whatever the methods used to document policies and practices the end result should be a coherent and consistent set of documents that cause the right things to occur.

- Vague statements do not signal intentions and, whilst easy to write, will result in variable methods where variation is detrimental to system effectiveness.

- Decide on the standards and practices before you start, otherwise you will end up having to rewrite documents later when you have learnt the disadvantages of being too flexible.

- Quality manuals that mirror the standard are a waste of paper – match the manual to your business so that it will be consulted.

- Any statement made by management at any level which is designed to constrain the actions and decisions of those it affects is a policy.

- A procedure is not a document but a sequence of steps to execute a routine activity. When documented, it may be five lines, five pages or 50 pages. Size is unimportant and is a characteristic of the complexity of the activity.

- A document may consist of one procedure or 50 and the level at which the procedure acts in an organization doesn't alter the fact that it remains a procedure. Quantity is unimportant and is a characteristic of the complexity of the organization.

- Many quality system documents are not procedures – they do not define how to proceed but they are nevertheless part of the documented quality system.

- Procedures should define who does what, when, where and how. Standards should define the acceptance criteria. Policies should define the boundary conditions. Records should define the results.

## Documentation task list

1  Prepare the document development plan.

2  Determine the documentation method.

3  Prepare a specification defining features you require of the documentation tools.

4  Set up the documentation tools (word processors, database etc.)

5  Provide user instructions for the document database.

6  Test the tools to verify they meet your specification before production use.

7  Determine the documentation standards.

8  Develop and issue a glossary of terms.

9  Produce the document development procedure.

10 Produce the document change procedure.

11 Nominate the authors.

12 Set up review panel for documents.

13 Determine approval authorities.

14 Issue the document requirements.

15 Issue the document-development procedures.

16 Create the control procedures.

17 Create the Policy Manual.

18 Review and approve the Policy Manual.

19 Review and approve the control procedures.

20 Issue the document change procedure.

21 Develop the supporting documents.

22 Review and approve the support documents.

23 Update the development plan.

## Documentation questionnaire

1. Have you decided on a paper system or an electronic system?
2. Have you tested the tools to verify they provide all the features you require?
3. Have you provided access control to the database?
4. Have you decided how the documents will look on screen and on paper?
5. Have you defined the numbering system?
6. Have you decided how document approval will be indicated?
7. Have you decided on the document dating convention?
8. Have you chosen the document revision convention?
9. Have you decided how you will indicate revisions in draft documents?
10. Have you chosen a style for the documents and communicated it to the authors?
11. Have you decided what to do about existing stocks of paper forms?
12. Have you decided what to do about existing documents produced using different software packages?
13. Do you know which documents you need to produce?
14. Have the authors, reviewers and approvers been nominated?
15. Do the authors know how to operate the tools?
16. Have you chosen the location for the quality system documents?
17. Have you decided what to do about electronic versions of documents located on remote servers or computers?
18. Have you decided what to do about the control of derived documents?
19. Have you defined the external documents and what to do about their control?
20. Have you provided a document in which special terms can be defined?

21  Have you defined what a 'Policy' is and provided guidance on their preparation?

22  Have you defined what a 'Procedure' is and provided guidance on their preparation?

23  Which version is to be the master: the paper version or the electronic version?

24  Will the forms be included in the procedure or be separate documents?

25  Have you decided how you will provide immunity to organizational change?

26  Have you decided which procedures are to be common to all sites?

27  Are all control procedures traceable to policies?

28  Are the policies compatible with the control procedures?

29  Are all supporting documents traceable to the control procedures?

30  Are all forms traceable to the procedures that requires their use?

31  Have you decided what will constitute your Quality Manual?

# Annex A
## Specimen procedure in flowchart form

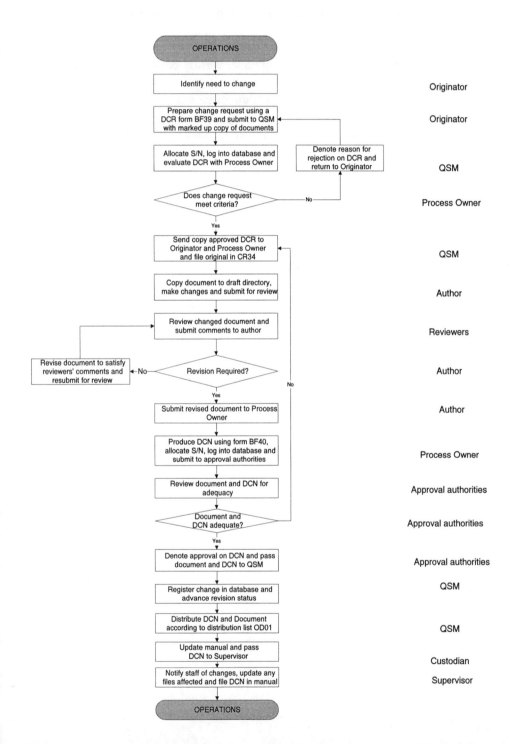

# *Annex B*
# Specimen documentation standard

## 1 PURPOSE

This document defines the presentation requirements for documents that form part of the management system.

## 2 REQUIREMENTS

### 2.1 Titles

The title of a document or the heading of a paragraph or section should reflect, as closely as possible, its scope and purpose. The title may be used for retrieval purposes so it needs to be accurate whilst being brief. System documents tend to describe tasks and activities therefore the title should either be a task or an activity (for example, Auditing, Planning etc.) It is not necessary to include the word 'Procedure' in the title.

### 2.2 Expressions

The word 'shall' is to be used to express provisions that are binding and 'should' to express a provision that is optional.

### 2.3 Abbreviations

The first time an abbreviation is used in text it should be spelled out in full and be placed in parentheses following the term used. Common abbreviations shall be added to ST01.

### 2.4 Heading numbering

Only headings shall be numbered. Decimal section numbering shall be used up to four levels (that is, 3.2.1.1). Thereafter sub-headings shall be by letter (that is, a), b), c) etc.) Any further divisions shall be Roman numerals (that is, i), ii), iii) etc.)

### 2.5 Style sheet settings

#### 2.5.1 Templates

System documents shall be prepared in accordance with OP32 in MS Word 6 using Template Cproc.dot, Std.dot or Oproc.dot, the characteristics of which are defined below. ODs and CRs shall be developed in a style to suit their purpose.

#### 2.5.2 Font

| Heading 1 | Headings 2 and 3 | Heading 4 | Body text | Footers |
|---|---|---|---|---|
| Arial 14 bold | Arial 12 bold | Arial 12 bold italic | Arial 12 regular | Arial 12 regular |

### 2.5.3 Paragraph

Before 0pt

After 6pt

Single

Justified

### 2.5.4 Margins

| Top | Bottom | Left | Right | Gutter | Header | Footer |
|-----|--------|------|-------|--------|--------|--------|
| 1cm | 2cm | 2.5cm | 2cm | 0cm | 1.27cm | 1.27cm |

### 2.5.5 Page size

A4 (210cm x 297cm)

### 2.5.6 Layout

Section start new page

No difference between odd and even pages or first and second pages.

## 2.6 File names and directories

A directory structure shall be established and maintained on the LAN in accordance with OP32 for all management system documents. Draft documents shall be saved to a DRAFT directory and when approved the documents shall be copied to an ISSUED directory. Subdirectories shall be created for each class of document (see below).

File names shall be the same as the document number and version identity shall be denoted by -A or -1 depending whether it is a draft or an approved version. Previous versions of documents shall be moved to an ARCHIVE directory. The version identity shall be changed when it is decided that a new version of a document is printed for distribution.

## 2.7 Dating conventions

CPs and OPs shall be identified with the SAVE DATE in the form dd-MMM-yy.

Form numbers and date shall be added manually to the form as it may need to be saved under another file name when details are added.

Standards, Output documents and Notices shall be identified with the SAVE DATE in the form dd/mm/yy in the footer right corner.

## 2.8 Issue conventions

Documents issued for review purposes shall bear letter issue notation. Approved quality system documents shall bear number issue notation commencing at 1. The issue notation shall be associated with the document reference number such that if a document is given a new reference number, the issue status will revert to 1.

CPs, OPs and operational STs shall have the issue status displayed in the heading block.

The convention for denoting the issue of reference STs, ODs, forms, labels and notices shall be AAXX-X, where X is the issue number (for example, OD06-1 is issue 1 of OD06). The issue status shall be displayed in the footer above the date.

The issue notation of forms shall be associated with the form title, such that if a form title changes, the issue status will revert to 1.

Documents that are not issued or accessible to staff other than the user need not carry any issue status.

## 2.9 Denoting approval

Approval of documents may be denoted either on the document or on an associated Document Approval Record (BF15). All CPs, OPs and STs shall have a BF15.

## 2.10 Flowcharts

Flowcharts included in procedures should be placed as near to the front of the document as possible. The responsibilities for each action and decision should be added on the right adjacent to the box. The words inside the boxes may be reduced for clarity and brevity. The flowchart summarizes the procedure – it does not replace it. It is needed to provide understanding, especially where there are many interfaces and options that are difficult to describe unambiguously in the text. It is also needed to identify opportunities for improvement. Flowcharts shall be produced using Visio.

## 2.11 Forms and labels

Forms and labels shall not be included in the procedure. All forms and labels shall be contained in Parts 3 & 4 of Procedures Manual PD04. The page numbering shall not include the forms and labels. Forms shall be produced using MS Word or Excel Forms shall specify the form number, date and issue in the bottom right margin and the related document reference number in the lower left margin. The issue notation shall be of the form BF26-1, BF26-2 etc. Draft issues of forms and labels shall not be used. Where forms have more than one side, each side shall be annotated with the side number (for example, side 1 of 2; side 2 of 2). Labels shall be copied onto A4 size paper and the colour denoted when it is other than white.

## 2.12 Notices

All Notices shall form part of the management system since their observance should be essential for the effective operation of the system. Their control also enables each site to be set up on a common basis.

## 2.13 Tables and figures

Tables and figures, other than flowcharts and forms, should be included within the text of the document and created using MS Word or Excel.

# Annex C
# Specimen role description for a process owner

## ROLE

The role of the Process Owner is to design, develop and improve specified business processes that enable the organization to meet its corporate objectives.

During quality system development the Process Owner leads the process development effort and, following development, seeks continual improvement in process performance.

## RESPONSIBILITIES

Irrespective of other responsibilities, the Process Owner is responsible for:

1 Designing a specified business process (using the prescribed development tools) that coverts specified inputs into outputs to satisfy the quality system requirements and business objectives.

2 Defining policies that will govern the execution and performance of the process.

3 Identifying the documentation requirements for each of the documents needed to execute the process in terms that will ensure they fulfill their purpose and meet quality system requirements.

4 Selecting and appointing authors for the various documents and ensuring these authors deliver documentation according to the agreed requirements.

5 Managing the development of the documentation identified and ensuring its compatibility with interfacing processes.

6 Identifying resource requirements for the development and installation of the process and ensuring these resources are deployed effectively.

7 Conducting initial training of staff in the operation of new processes and the interfacing practices.

8 Conveying quality system requirements to staff and securing their commitment.

9 Managing the installation of new processes and ensuring their smooth transition into service.

10 Establishing performance metrics and measuring process performance.

11 Managing improvement programmes that will simplify practices, reduce process variation and optimize performance.

## AUTHORITY

The Process Owner has the delegated authority of the Quality System Manager to:

1 Determine the sequence of tasks required to make the process effective.

2 Assemble review and improvement teams for developing and improving the process.

3  Approve process documentation and authorize its release for use.

4  Change the process should improvement be necessary to meet the defined goals.

5  Cause compliance with the requirements for the effective operation of the process.

## ACCOUNTABILITY

The Process Owner is accountable to the Quality System Manager for:

1  Putting in place a business process that satisfies the needs of the business and the quality system requirements.

2  Meeting target dates for the completion of process development and approved improvement programs.

3  The communication of policies and practices to staff operating the process.

4  The performance of the process in meeting client requirements and organizational goals.

## QUALIFICATIONS

The Process Owner shall possess the following qualifications.

### Position

The Process Owner shall be a person holding a position in the organization that is either equal to or above all of the staff who contribute to the process.

The Process Owner shall be a person whose other duties do not create conflict in terms of availability or priorities.

### Knowledge and experience

The Process Owner shall have a knowledge and experience of:

1  ISO 9000

2  The business of which the process is a part

3  The process for which he/she is the designated Process Owner

### Training

The Process Owner should have the following skills and experience:

1  Leadership

2  Quality system design, development and implementation

3  Planning, conducting and reporting audits

4  Interpersonal and assertiveness

5  Oral and written communication

6  Personnel management

7  Change management skills

8  Time management

## Personal characteristics

The Process Owner should be:

1  Inquisitive, so as to seek out information naturally.

2  Systematic, so as to search for information in a logical and progressive manner.

3  Assertive, so as to obtain information fairly and effectively.

4  Analytical, so as to derive proper conclusions from collected facts.

5  Objective, so as to maintain a focus on the objective of the process.

6  Friendly, so as to create a co-operative relationship with the development team.

7  Sensitive, so as to react effectively to differences of opinion.

## Chapter 7

## Document and data control

### Document control strategies

Document control is a process for controlling the development, use, maintenance and disposal of documents used and generated in the quality system. Chapter 5 identified the documentation needed to build the system and Chapter 6 dealt with the development of these documents. The various tools and procedures that can be used for documentation development were identified but not described in any detail. These aspects, and the controls required following release of documents for use, are the subject of this chapter. It is interesting to note that ISO 9001 does not impose any controls over the development of documentation directly, except prior to their release. The standard requires documents and data to be reviewed and approved for adequacy prior to issue. However, a closer examination will reveal that:

- The responsibility for producing documents has to be defined and documented (clause 4.1.2.1) since it is 'performing work' that affects quality.

- Trained personnel have to be assigned to produce the documents (clause 4.1.2.2).

- Adequate resources have to be provided for producing and maintaining documents (clause 4.1.2.2).

- The steps needed to deal with any problems requiring preventive action are required to be determined (clause 4.14.3b).

There are, therefore, plenty of reasons for controlling the development of documents as well as their use and maintenance.

There are many software tools on the market that can be used for developing and controlling documentation. Some allow you to import your documents into a database and others merely handle the data and leave the documents in their original form. You can also choose a proprietary solution or design your own using a proprietary database such as Microsoft Access, Oracle or Lotus Notes. However, Lotus Notes is not a relational database, unlike the other two, and therefore has no query ability.

What do you look for in a software tool? It depends on what you want to do now and in the future. Large organizations will probably require a network solution. Small organizations may well operate adequately with a standalone PC solution. Before you begin, you need to consider a number of options.

### On-line or paper-based solution

If most employees use computers to aid them carry out their activities, the keyboard skills required will be widespread and therefore a computer-based solution may be appropriate. Even if the computers are used to aid manual work (such as with process plant operators), a computer-based approach may be a better solution in order to remove paper from a dirty environment.

If most employees perform work without the aid of computers, then the keyboard skills will not be widespread in the workplace and so a paper-based approach is likely to be more appropriate.

### Computerized controls of printed documents

If you opt for an on-line approach, it will follow that data controls will also be computerized. If, however, you opt for a paper-based solution, you can use a word processor to produce the documents and a database to control the related data and print the documents for distribution. What you need to decide is which is to be the master: the paper version or the electronic version. If you don't have access and change controls over the electronic version, then it may be wiser to designate the paper version as the master (that is, the version which is accurate and under change control).

## Database design

### Number of databases

One database for holding all the quality system documentation would be preferable but may not be possible. The quality system documentation is relatively static and does not change from project to project, product to product, or contract to contract. Job-specific documentation should be kept separately from quality system documentation, so there will be many storage locations for the data. However, at one or more levels in the quality system, procedures will become focused on functions, locations, technologies etc. and may be in a form that is incompatible with newer documentation. It may not be possible to move these documents into one database and so allowance should be made for controlling more than one database with perhaps different control procedures.

## Database specification

Designing a new database should be treated as you would the design of a new product. You should prepare a specification that defines what you require of the database. You may not get it right first time but every time you change the design you should update the specification so that you have a description of the features provided. This document will become a point of reference for future changes and serve to prevent changes being made that may not work or will cause problems with the information in the database.

## Choice of software

One aspect you need to consider in selection of the software is the number of users and the existing software that is used. If you don't have any common database tools in use, the number of licensed copies needs to be decided. You also need to take into account how you will deal with upgrades. It can be very costly to settle on one product and then find it is not supported or that it has to be installed on every workstation because network versions are not available. If you use databases for other jobs, it may be more practical to choose a solution that will work with existing software. Take care to establish the versions of the software used as this may give rise to compatibility problems.

Some of the features you may wish the database to possess are covered in the following sections.

## Document views

You will need to decide how you want documents to be viewed on screen. It is not merely a case of having read-only views. Developers need to view documents in preparation as well as those approved and obsolete. Users need access to approved documents only. In addition, users need to retrieve documents quickly for reading and you should not assume that users will know the document number or title. They may need to search by document type, process, function, location or technology. The software needs to provide the flexibility for you to change views to suit your needs and to select various sorting options.

## Document attributes

When listing documents, you need some basic information on the index screen:

- Document type (CP, OP, GD etc.)
- Document reference number
- Document title
- Document status (draft/approved/obsolete)
- Date of creation (useful to indicate the age of the document)
- Date of release

Reference numbers may not be critical if titles are unique. In addition, grouping by document type, process, function, location or technology may be desirable.

## Access controls

You should appoint a Database Manager to control database design, maintenance and access controls. Various people will need access to the database apart from the Database Manager:

- Authors need 'read and write' access
- Reviewers need 'read and comment' access
- Approvers need 'read and approval' access
- Users need 'read-only' access

You may also need to limit access to some types of documents, such as contracts, confidential documents or certain operating instructions. (You would not want the operating instruction for disabling the security system to be accessible to all!)

## Drafting controls

The database should provide the ability to either draft documents using any word processor and then import them or create the document with the built-in word processor. Lotus Notes has this facility but the word processor does not have the features of standalone word processors such as MS Word or Word Perfect.

If you create documents in a separate word processor, the headers and footers and the formatting may not be fully importable. Importing can provide a secure means of preventing change as the document can be given write-protection and 'read-only ability'. However, in the drafting stage, it can be irritating to have to import a new version every time you send out a draft for comment; people sometimes forget to change the revision status, which then provides more than one document of the same revision.

A database with word-processing features, such as Lotus Notes, provides users with the ability to change documents in the database without changing the original in the separate word processor. It needs a strict discipline to keep both versions the same unless you link the files. If you link files you have to place both files on the same server. Import a document from a standalone computer and you lose control of that version unless you embed it without links. This can cause problems too as flowcharting tools may not be provided. If you embed documents as icons, then beware of the launching time (the time for the hour glass to disappear). The larger the document the longer it takes to launch and this can be irritating to browsers.

## Reviewing documents

An on-line system needs to provide for draft documents to be distributed for review by a Review Panel. The database should allow you to group reviewers from a controlled list so that when you send out a new draft you can send it to the same group of reviewers.

On receipt of the draft, the reviewers need read access plus the ability to add comments. This can be achieved either by attaching notes or by a comment column on the document.

Following review, the reviewers need the ability to return the commented document to the author in a manner that is instantly recognizable to the author, otherwise the author may think he/she has several versions of the same document and delete a reviewer's comments by mistake.

## Revision controls

When documents are being drafted you need to employ a revision convention that indicates the documents are not approved, such as *letter revision status*. When approved you should use a *numerical revision status*.

Some packages automatically advance the revision status by one each time you revise the document. It can be irritating if your revision was not complete when you last saved the document. You need the ability to advance the revision not when you save a document but when you release a document that is different from the previous version.

## Approval controls

When a document has been reviewed and the reviewers are content that it meets the requirement, the database needs to provide the author with the ability to release the document to the approvers. In many cases, there may be only one approver, but the tool needs to provide for more than one approver.

On receipt of a document for approval, there should be provision for the approver to indicate approval in such a way that his/her name and date of approval is entered into the document in an unchangeable field.

If the approver does not approve the document, provision needs to be made for the approver to disapprove or to deny approval and for reasons for denial to be entered.

## Database testing

A new database should not be released for use until you have proven that it is fit for purpose. Test the features to verify that they function correctly with the necessary security measures in place. Try doing things that should not be possible in order to prove that the design is robust.

Always keep a prototype that is not released and which is used to test changes before releasing them to users. Don't change the released version until you have proven that the changes work correctly, otherwise you may create difficult problems that are time-consuming to fix later. Trying to design a database when data is being input by staff will significantly extend the development time.

If you have designed a database using the latest version of the software, verify that it can be used with all earlier versions that users have on their PCs. Obviously it would be wiser for everyone to have the same version of the software but with a large user population, this may not be practical or cost-effective.

## Programme impact

Should you choose a database solution to control your documents you need to consider the programme impact of this decision. The procurement and installation of new tools takes time. You also need to allow time for user training. If tools are already used in the organization, training in their use may be less than if the tools are new. Staff will, however, need training in the use of the specific solution you have designed or procured. If you go for a custom design then you need to allow far more time for its development. It can take several months to get a new database up and running without problems. A do-it-yourself solution is always fraught with problems if you don't have the necessary skills.

## Document control requirements

The following list summarizes the document and data control requirements of ISO 9001 rephrased as instructions.

## General

1. Establish and maintain document control procedures for documents of internal origin.
2. Control all documents that relate to the requirements of the standard.
3. Establish and maintain data control procedures.
4. Control all data that relates to the requirements of the standard.
5. Establish and maintain document control procedures for documents of external origin.
6. Control all documents of external origin that relate to the requirements of the standard.

## Document and data approval and issue

1. Review documents for adequacy prior to release.
2. Review data for adequacy prior to release.
3. Ensure that documents are approved by authorized personnel prior to release.
4. Ensure that data are approved by authorized personnel prior to release.
5. Establish and maintain a master list or document control procedure for identifying the current revision status of documents.
6. Ensure that the master list or document control procedure is readily available.
7. Ensure that the pertinent issues of appropriate documents are available at all essential locations.
8. Prevent unintended use of invalid and/or obsolete documents.
9. Identify any retained obsolete documents.

## Document and data changes

1. Ensure that changes to documents are reviewed by the same functions that performed the original review.
2. Ensure that changes to documents are approved by the same functions that performed the original approval.
3. Provide the designated functions with access to pertinent background information upon which to base their review and approval.
4. Identify the nature of change in the document or appropriate attachment.

The sections that follow provide suitable solutions to these requirements for both paper-based and electronic systems. A comprehensive guide to these requirements is given in the *ISO 9000 Quality Systems Handbook*[1].

## Document development procedures

An example of a document development procedure is given in Annex A. This is a procedure for use in a paper-based system. ISO 9001 does not require a procedure for document development and if you study the requirements carefully you will learn that they only address the release and change of documents. Their purpose is to prevent use of invalid documents. Invalid document are those that are:

- Documents of the wrong issue status for a particular task
- Draft documents that have not been destroyed
- Documents that have not been kept up-to-date with amendments
- Documents that have been altered or changed without authorization
- Copies of documents that have not been authenticated
- Unauthorized documents or documents not traceable through the quality system
- Illegal documents

How you produce the documents referred to in ISO 9001 is up to you. Consider what would happen, however, if documents varied in their layout, style, terminology, depth and presentation. Staff would probably be confused and query the difference, which might result in significant inconsistency in the degree of control employed. One way to overcome this is to provide criteria for reviewers and approvers so that documents passing through this process are consistent. If you only apply these standards at the 'inspection' point then the authors will become unhappy with the feedback and complain that you should have told them what you wanted before they started. Hence, the need for planning. To prevent this type of problem, you need *document standards*.

If you don't define the development process, you may inadvertently cause inconsistency in the drafting and review processes, costing you time. Annex B of Chapter 6 has an example of a document standard. Authors may use the wrong tools, route the drafts to the wrong people and, most important of all, produce documents for which there is no requirement! To prevent this type of problem, you need a *document development procedure*.

You can embody both procedure and standard in one document and it does not have to be an 'i' dotting and 't' crossing document. The document development procedure may

---

[1] *ISO 9000 Quality Systems Handbook*, David Hoyle (Butterworth-Heinemann, 2nd Edition, 1994).

differ for different types of document but the elements should be the same. Procedures for developing contracts, specifications, drawings etc. may need to be different because the users concerned are different and the nature of the document is different from procedural documents. Drawings, for instance, may be single sheets and not multi-page documents. Handbooks may be contained in ring binders with drawings, pull-out sections and dividers. Contracts will be in a legal style with signatures at the back instead of the front.

## Elements of document development

### Identification

Documents may be identified by number or title. When numbering documents the number may or may not include a document code to indicate its type. As the document controls vary with type of document, coding is a means of identifying which controls are to apply. You exercise development and change control over policies and practices but not records (except preventing change) and your control over external documents is limited to their distribution, use and prevention of change. The types of derived document are so varied that the controls also vary. By assigning codes to documents (see Chapter 6) you can designate which controls are to be applied to which document.

### Preparation

The documents to be developed should have been identified during system design. A development plan should be produced which assigns an author and a target date. Authors should be given training before commencing preparation to ensure that they understand:

- How to use development tools
- The terminology to be used
- The standards to be applied
- The process to follow

### Review

Reviewers should be nominated beforehand so that authors know who they have to satisfy when preparing the document. Reviewers also need to be trained, otherwise they may give conflicting advice. Reviewers need the same training as authors. Reviewers need not be given document-development skills. Review criteria should be established so that reviewers know what they need to do. Many a problem has been created by reviewers employing different standards. Some read documents and look only at what is written – never asking themselves what is missing. Others only correct the spelling and grammar – never giving you the benefit of their experience. Reviewers should be looking at the statements and asking themselves questions such as:

- Where does this come from?

- Who does this go to?
- Where does this go to?
- What happens if this is incorrect?
- What standard does this need to satisfy?
- Who checks this?
- Is this auditable?

Why do you submit documents for review? If you are the only user of the document, its review is not necessary. If the document is to be used by others then they have a role to play in ensuring that the document is fit for purpose. You may argue that the review is carried out by the approver, and so it is – but unless the approver is the only other one who will use the document, this is insufficient. The approver may be the manager but not the user. The manager may not know the detail necessary to judge whether it is fit for use. If the people using the document have not had a hand in its development, they will not feel they own it. Ownership is key to gaining commitment. You want the manager to own it, so you obtain his/her approval. To get the user to own the document, ask him/her to elect a representative to review the document and then, if the document is unsuitable after review and approval, the users know that you were not at fault. You gave the users the opportunity to contribute and they failed to do so effectively.

### *Approval*

Approvers should also be nominated beforehand. Approvers need to be trained, otherwise they too may give conflicting advice and they need the same training as the reviewers. Approval criteria should be established so that approvers know what they need to do. The approver's task is on a higher level than that of a reviewer. Approvers are carrying a responsibility and have authority to approve a document on behalf of the organization. Approvers need to verify that the document:

- Is fit for purpose
- Meets its objective
- Implements the relevant policies
- Complies with requirements
- Matches the interfaces
- Reflects best practice

### *Release*

There are two forms of release: release for comment and release for use. *Release for comment* follows when the author has produced the first and subsequent drafts. *Release for use* follows when the approver has signalled approval to the first and subsequent versions.

It is important that these two releases carry an identification that readily distinguishes between them.

With paper systems you will need to maintain one or more lists that indicate the revision status. A document passes through three maturity levels: Draft, Approved, Obsolete. Your list should have two columns: one for Maturity and one for Revision. The Maturity column should indicate whether the document is Draft, Approved or Obsolete and the Revision column should give the pertinent revision of the document. Alternatively, you can choose not to display the Maturity level and indicate only the Approved version. The *pertinent* issue may not be the current or the latest. If earlier versions need to be used, there needs to be a method for designating their applicability. One such method is to designate the pertinent revision of a document within the instructions or specifications and direct the user to these documents when assigning the task.

### Distribution

Distribution occurs following release. With an electronic system, distribution can be made simultaneously with release. With a paper system, you will need distribution lists to indicate who gets what. Either way, a distribution network needs to be set up so that documents are available at locations essential to the performance of the tasks to which they refer. Distribution lists should be controlled documents to prevent documents being sent to the wrong people or the wrong locations. The difficulty arises when people move. This is particularly the case in large organizations. You need a process that will ensure that distribution lists are changed when there is a change in the organization. You need to accommodate:

- Changes in position, department and location
- New entrants and retirements
- Mergers, acquisitions and downsizing

There is no simple solution. You can use staff lists or a company personnel database but you need to establish its accuracy and how it is maintained. Many systems rely on the individual informing head office, which of course they don't – so the lists are often an inaccurate source of data.

One solution is to distribute documents to fixed locations under the control of a local document custodian where they perform the local distribution. With an electronic solution, the documentation resides on a server that is accessible to all so local distribution is less likely to be a problem. Knowledge of staff changes is only necessary for the review and approval panels where you have designated specific roles.

## Document change procedures

An example of a document change procedure is given in Annex B. This is a procedure for use in a paper-based system. Control of document change is necessary for two reasons: to prevent unauthorized changes being made and to cause desirable changes to be made.

With paper systems, unauthorized changes can arise when staff write on controlled copies of documents without authorization. You can make authorized changes to copies through *document change notices*.

Unauthorized changes can be made to electronic documents if the access controls do not prohibit 'write and delete access' to designated personnel. The principle difference with electronic versions is that unless the right access control is installed, unauthorized changes can be made to the original: a situation that is unlikely with a paper-based system as the originals should be locked away and only brought out when a request for change has been approved.

As with document development, you can limit the change procedure to a few simple steps: change document, review document, approve document, release document. However, to deal with problems requiring preventive action (clause 4.14.3 of ISO 9001) you would be wise to consider a few extra steps. Change control applies to all your documents, not only to the policies and procedures. It applies to contracts, specifications, drawings, handbooks and all those other documents that are used to produce products and services (except, of course, records).

### Change requests

*Change requests* provide a method of enabling personnel unauthorized to change a document to initiate a change to an approved document. So that you keep track of all changes, those with authority to change a document should also use the same procedure. Change requests can be by informal or formal means. A memo provides an informal means for requesting a change but doesn't have the constraints that will ensure the right information is given each time. You may wish to prevent:

- Ambiguity over which document is to be changed
- Ambiguity over what is to be changed
- Uncertainty about the reason for change
- Lack of information about the effects of change
- Processing multiple requests for the same change
- Work proceeding on a change before it has been evaluated

The way to overcome these problems is to use a Change Request Form that requires certain data to be provided by the originator. Give each request a serial number and use priority classification to schedule the implementation of changes. A typical document change request is given in Annex C.

## Change accounting

In order to cause desirable change and prevent undesirable change to documents you need some means of knowing:

- What changes have been requested
- Which requests have been approved
- Which changes are being worked on
- Which changes are awaiting review
- Which changes are awaiting approval
- Which changes have been approved
- Which versions have been released

With an on-line forms package the database can provide this history by sorting the individual change requests providing the change request has data entered in such fields.

With a paper-based system, a separate register of changes can be generated with columns for entering the date on which the change passes through each stage in the process. A manual sort can be tedious so a spreadsheet can be used to automate this task. The only difference is that the data needs to be entered separately as it will not be automatically extracted from the request.

## Change evaluation

You can react to every request and get yourself in a tangle – making changes one day then changing them back again the next. The change needs to go through as rigorous a process as the original document, otherwise you may end up with a document that is unfit for purpose. Every change ought to be subject to evaluation. The evaluation of the change request does not have to be by the original document reviewers or the approvers. The evaluator needs to apply the same rules and common sense as would be expected from the reviewers and approvers. The change evaluator might be a person or a panel depending on the significance of the requested change. If additional resources are needed to implement the change, then it would be wise to obtain the go-ahead first. There is little point in proceeding with a change if the potential costs will not be met. Often the costs cannot be determined by the change request originator and in such cases the request should pass through two stages: one to agree the change in principle and another to estimate the cost. Stage one of the evaluation is a stage for evaluating the technical feasibility and desirability of the change and should answer these questions:

- Can we make this change or does it conflict with our objectives? (Objectives can be meeting customer requirements, satisfying business goals, meeting ISO 9001 requirements etc.)

- Is this a change we have to make and if it is when does it have to be implemented?

- What are the benefits of this change; what will it improve?

- What adverse effect will this change have on related products, processes, services or documents?

- What are the consequences of not making this change?

- Should we therefore make this change?

- Should we release the change when complete or hold it and release it at a later date?

If the change is merely cosmetic then there is little point going through this process. You may therefore need a bypass loop for changes that do not affect the operation of product, process or the quality system (see below). A simple solution is to classify changes as either major or minor. Major changes are those affecting product, process or the system and minor changes are all others.

The change request should provide for the results of the evaluation stage to be passed to the originator and for the instructions to change to be passed to the author.

## Organizing change evaluation

The organization required to process changes needs some consideration. There are, as described in the previous section, several stages through which a change should pass depending on its magnitude. Processing every change through a review panel regardless of its impact may impose an undue administrative burden on the panel. Processing every change through one person may not only impose an undue administrative burden on that person but also result in ill-considered views and, hence, biased judgements. Passing each change through local management may result in the full impact of the change not being addressed.

A hierarchy is therefore necessary to make the process cost effective. There are various levels, each of which reflects the potential impact of the change.

**Level 1**   No impact beyond the document affected. Covers cosmetic changes and changes that do not affect practices, products or processes. These changes should be evaluated by the document owner.

**Level 2**   Impacts practices but confined to the document in which the practices are defined. These changes should be evaluated by the document owner and the functional managers affected by the change.

**Level 3**   Impacts products but confined to the product documentation. These changes should be evaluated by a Product Change Control Board.

**Level 4**   Impacts one process but confined to the documentation which describes the process. Such a change could impact all products using this process. These changes should be evaluated by the process owner and the Product Managers affected.

**Level 5**   Impacts more than one process, causing redesign of interfacing processes. These changes should be evaluated by a System Change Board comprising all the process owners.

**Level 6**   Impacts the system design such that major redesign is necessary. These changes should be evaluated by a System Change Board with representation from the System Review Board.

To make this work, you will need to appoint change controllers: that is, someone to whom the changes are routed and who decides the level to which the change request should be passed. Product development teams need their own change controllers and Change Control Board. In some cases, these can become quite complex where customers and subcontractors are involved[2]. With the quality system, you will also need a change controller and this is one of the functions of the System Development Manager. The roles of the System Development Manager, System Change Boards and System Review Boards are covered in Chapter 8.

### Change initiation

#### *Options*

Changes to documents can be made in two ways: by making changes to the document or by issuing a change instruction to change a copy of the document. You can leave the original unchanged or change the original without releasing the new version.

[2] See ISO 10007 on 'Configuration Management'.

With on-line documents you can make a single change to the original and, following its approval, you can send the new version out to all users at the press of a button (providing all users have access to the network).

With paper documents, change is not so simple. You have to change the original, obtain approval, obtain copies and then send these out to the users. This is the most frequently used method. However, there may be circumstances when this is not practical:

- The change may be so simple, it does not warrant the cost of reprinting the document.

- The change may have such an effect on other documents that the cost is prohibitive.

The 1987 version of the standard required that documents be re-issued after a practical number of changes have been made but this provision has been removed. The requirement stems from the days before word processing when changes were promulgated by amendment leaflet or change notes and one had to paste additional paragraphs over ones which were crossed out. In such circumstances there were only so many changes of this nature you could make before the document became unusable and a potential source of error. If you still operate in this fashion, the number of changes may well be a limiting factor.

However, there are practical reasons why, even in the IT age, it may not be prudent to reissue a document after each change.

There are several types of changes you may need to consider:

- Changes affecting a whole range of documents
- Changes affecting many pages of a single document
- Changes affecting a few pages of a single document

**Changing the original**
On receipt of instructions to change a document, the nominated author should process the change using the established document development procedures. However, prior to doing so, you may wish to consider how the changes will be brought to the attention of the document users as it may affect how the changes are made. There are several methods available:

- Sidelining
- Underlining and strike through
- Emboldening
- Change Notice

Some word processors provide for marking revisions to documents by sidelining, underlining and strike through. They can be switched off and on at will. Whilst these features are useful in drafting and reviewing documents they are unsuitable for general use. When you accept all the changes, the formatting may not be perfect and further revision may be needed. Sidelining can be used to indicate paragraphs that have changed but are inaccurate for telling readers what has been deleted.

Emboldening or using colour with on-line documents has the advantage of indicating changed text and can be removed without causing a material change to the document.

The only method that allows you to explain the changes is the change notice issued with the changed document. A compromise is to use the revision features of the word processor whilst in the drafting mode and then use the change notice following approval.

With on-line systems, you need to ensure that the original remains protected and that the changes are either made on a copy or a new version is created.

### Document re-issue
Where a change affects many pages the document should be re-issued. Even if the substantive change is minor, the knock on effect in double-sided documents with diagrams etc. can be to change every page. With modern word-processing techniques, even adding a full stop can throw out several pages.

### Page changes
Where a change affects only a few pages you can issue the changed pages with an amendment instruction informing the recipient of which pages to change. Alternatively you can use the Document Change Notice (DCN) to add new pages and amend text at the same time. If only a few words or figures are affected, then the DCN is by far the least expensive and the quickest method.

### Holding changes
As an alternative to actually issuing changes, you may wish to process the change requests to the master and hold re-issue of the document until a suitable number of changes have been made or a significant proportion of the document has been changed. It is not the number of changes that is significant – a single change could have far greater effect than 20 minor changes. With small documents, say three to six pages, it is often easier to re-issue the whole document for each change.

### Global change
For the change that affects a whole range of documents you will either have to re-issue the complete set of documents or employ a Global Change Notice. When the cost and time required to process a change that affects many documents is prohibitive then something like a Global Change Notice (GCN) is a useful tool to have in your quality system. With a GCN you can instruct document holders to make changes to certain documents

in their possession without having to identify every document. For example, if a component code number changes, a GCN can authorize the code to be changed in any documents which specify that component without having to process hundreds of documents. When the document is subsequently revised for other reasons, the GCN can be embodied so that over a period of time all documents will eventually be brought up-to-date. You will need a means of alerting staff to the issue of a GCN but, if you control your distribution lists, this should not present a problem. If all your documents are electronic, there may be no need for a GCN mechanism, except for classifying a change.

### Change notation

Dates can provide a suitable means for indicating revision status but, during the revision process, a date may not be accurate enough and can be misleading. A better method is to use letter revision during drafting and numerical revision for approved documents. When changing an approved document, to indicate various drafts you can use the convention 1a, 1b, 1c etc. so that when approved 1c becomes revision 2.

### Change review and approval

ISO 9001 requires changes to be reviewed by the same functions or organizations that approved the original. The reason for this is to prevent different criteria being applied and to ensure the same level of expertise is applied. If other functions approved the document, there is a possibility that they would not apply the same judgement and, as a result, adversely affect product quality. This is not to say that others may not have the same qualifications – under certain circumstances, provisions can be made to permit other suitably qualified personnel to review and approve changes to documents.

### Release

When the changed document has received approval it is ready for release. However, when releasing a new revision of an approved document, you need to inform the users what has changed and why the changes were made. For this purpose you can add notes to the document or attach a Change Notice. The former practice is suitable for on-line systems and the latter for paper-based systems. An example of a Change Notice is given in Annex D.

### Distribution

The changed document should be released to the same distribution as the original unless the changes affect functions that were not affected by the original.

## Document accessibility

The accessibility of on-line documents was dealt with above. With paper documents this creates a problem that requires different procedures.

In order to make sure that paper documents are available you should not keep them under lock and key unless you want to give keys to all users as you do, in effect, with an on-line system. You need to establish who wants which documents and when they need them. If there is a need for access out of normal working hours then access has to be provided. The more copies there are, the greater the chance that documents will not be maintained so minimize the number of copies. A common practice is to issue documents to managers and not the users. This is particularly true of quality system documents. One finds that only the managers hold copies of the Quality Manual. In some firms all the managers reside in the same building, even along the same corridor, and it is in such circumstances that one finds that these copies have not been maintained. It is therefore impractical to have all the copies of the Quality Manual in one place. Distribute the documents by location not individual. Distribute to libraries or document-control centres so that access is provided to everyone and so that someone has the job of keeping them up to date.

## Obsolete documents

It is unnecessary to remove invalid or obsolete documents if you provide staff with the means of determining which pertinent issues of documents to use. There are often valid reasons for retaining obsolete documents. What may be obsolete in one situation may not be obsolete in another. In simple terms an obsolete document is one which is no longer required for operational purposes. One may need to remove copies of previous versions of a document but retain the master for reference purposes. You cannot demonstrate to someone that you corrected a deficiency if you don't retain the version that contained the deficiency as well as the subsequent version. If you do not have a means of readily distinguishing the correct version of a document, amendment instructions should require that the version being replaced is destroyed or returned to the document controller. If you allow uncontrolled copies to be taken, removal of obsolete documents becomes more difficult. However, providing you have a means of distinguishing controlled and uncontrolled documents you should have no problem. If there is no means of determining current versions, the chances of using the wrong document are significantly increased if several versions are accessible at the same location.

With an on-line system, obsolete documents can still exist. There is a higher probability that they will exist because it is so easy to create a new version and forget to delete the previous one. Your document control tools should therefore automatically move previous versions to an archive.

One way of identifying obsolete documents is to write 'SUPERSEDED' or 'OBSOLETE' on the front cover but doing this requires that the custodian is informed. When a version of a document is rendered obsolete by replacement with a new version then its withdrawal can be accomplished in the amendment instructions that accompany the revision. When documents are rendered obsolete by their total replacement then this can also be accomplished with the amendment instruction. However, where a document becomes obsolete and is not replaced there needs to be a Document Withdrawal Notice (DWN) which informs the custodian of the action to be taken and the reason for withdrawal. An alternative is to process a change notice to the relevant list of documents or the document which invokes the document to be withdrawn.

There are various reasons why a document may become invalid. With paper versions of documents there is no simple way of indicating their validity. By printing authentic documents on coloured paper, or providing paper with a special header, one can detect unauthorized copies. Placing the approval signatures on the front sheet will immediately identify an approved document. Adding the total number of pages of the document to each page will enable you to detect whether the document in your possession is compete. However, the onus must rest with the user who, if properly trained, will not use invalid documents.

## Control of external documents

You need to define what constitutes an external document. In principle, an external document is a document that was originated outside the quality system and remains under controls that are not part of the system. A document may be external to the quality system and internal to the division or external to the division and internal to the company The control which you exercise over external documents is somewhat limited. You cannot, for instance, control the revision of such documents; therefore all the requirements concerning document changes will not apply. You can, however, control the use and amendment of external documents. You can control use by specifying which versions of external documents are to be used and you can remove invalid or obsolete external documents from use or identify them so that users will know they are invalid or obsolete. You can control the amendment of external documents by controlled distribution of amendment instructions sent to you by the issuing agency.

There are two types of external documents, those in the public domain and those produced by specific customers. In some cases the issues of both types of documents are stated in the contract and therefore it is important to ensure that you possess the correct version before you commence work. Where the customer specifies the issue status of applicable public domain documents to be used, you need a means of preventing their withdrawal from use in the event that they are revised during the term of the contract. Where the issue status of applicable public domain documents is not specified you may

*Document and data control* **259**

either have a free choice as to the issue you use or, more likely, you may have to use the latest issue in force. Where this is the case you will need a means of being informed when such documents are revised so that you can obtain the latest version. The ISO 9000 series, for instance, is reviewed every five years and so may be revised at five-year intervals. With national and international legislation the situation is rather different as these can change at any time. You need some means of being alerted to changes which affect you and there are several options from which to choose:

- Subscribing to a standards updating service with the issuing agency

- Subscribing to a general publication which provides news of changes in standards and legislation

- Subscribing to a trade association which provides bulletins to its members on changes in the law and relevant standards

- Subscribing to the publications of the appropriate standards body or agency

- Subscribing to a society or professional institution which updates its members with news of changes in laws and standards

- Joining a business club which keeps its members informed of such matters

Also, as a registered company you will receive all kinds of complementary information from government agencies advising you of changes in legislation. As an ISO 9000 registered company you will receive bulletins from your certification body on matters affecting registration and you can subscribe to *ISO 9000 News* to obtain world-wide news of events and changes in the ISO 9000 arena.

The method you choose will depend on the number and diversity of external documents you need to maintain and the frequency of usage. An example of an External Document Control Procedure is given in Annex E.

## Controlling data

In some organizations data control is big business with large Information Systems (IS) Departments responsible for the applications software, networks, distributed technologies and the data processing operations[3]. In this section we will deal only with the fundamental concepts relative to ISO 9000 requirements.

---

[3] A very practical book on this topic is *Business Information Systems*, H D Cliford and A G Sutcliffe (Prentice Hall, 5th Edition, 1994).

## The nature of data

Data presents a special case and in the context of ISO 9000 the data referred to is data used in carrying out work. The previous sections addressed documents rather than data (although such documents may well contain data). Data is the representation of facts, concepts or instructions in a formalized manner suitable for communication, interpretation or processing by human or automatic means[4]. When you develop a document you will invariably collect data for use in its creation. If the data was originated by those who prepare, review or approve the document, the review and approval processes through which the document passes provide adequate control over the data. Problems arise when the data used originates outside the document development process. Data such as Data Sheets, Supplier Lists, Cross Reference Lists, Properties of Materials, Electronic Component Data Books etc. can be treated as documents unless they use as their data source some other document that is outside the control of the quality system. Such documents would therefore need to be treated as external documents.

Problems also arise when data used in one document is copied into another document. A typical example would be a flow diagram that is used in more than one document. You will need to designate which is the master document or separate the flow diagram and place it under document control. You would need to reference the source of the diagram so that in the event of a change being needed, the author retrieves the master for change and does not change the copy. It would not be practical to maintain a list of where an item of data is used as the owner would have no control over who uses the data and what it is used for.

The dangers from the use of invalid data are more profound with scientific, financial and engineering data. Care should be taken over sources to ensure the data is authentic and current before use. As a general rule, always reference the source of data so that its currency can be verified at any specific time.

## Database controls

Databases have been referred to in several chapters, either as repositories for documents or for data. They are the modern equivalent of the filing cabinet and exhibit some similarities:

- Access needs to be controlled otherwise data could be read by unauthorized personnel. With the filing cabinet, locks are provided and keys given to trusted staff. With the database, access is limited to those using the key passwords.

- Data location needs to be controlled otherwise it may not be retrievable when required. With the filing cabinet, wallets identified with suitable labels keep papers

---

[4] *BS 7799 Information Security Management* (British Standards Institution, 1995).

together and enable quick retrieval. With the database, data is stored in directories with suitable labels.

- Removal of data needs to be controlled to prevent modification and replacement by unauthorized personnel. With the filing cabinet, controlled data would be logged in and out using authorized signatories. With the database, editing and delete access is limited only to designated personnel using key passwords.

There are, however, some distinct differences between the filing cabinet and the database. If a distributed database is employed, data can be accessed, input, changed or deleted from remote locations at any time of the day or night. The logging-in system is automatic and audit trails are provided so that it is possible to discover who did what and when. The primary difference, however, is that a database is more than just a repository for data. The data can be manipulated to produce reports and many functions can use the same data for various applications. It is therefore necessary to ensure there is no redundant data otherwise discrepancies will in all probability appear between different data files.

As with the filing cabinet, a custodian (often called a Database Manager) is needed to maintain the database. The Database Manager should have responsibility for the assignment of access rights and for changes to the database design. This person need not be concerned with the actual contents of the database as these are the responsibility of the user departments. There should, therefore, be a Database Design Change Procedure so that the full impact of any changes is evaluated and tested before implementation.

## Network security

If the databases reside on computer servers or mainframe computers that are accessed through a local or wide-area network, there will be a need for a Network Manager. The Network Manager should have responsibility for the security of the network[5], its reliability and maintenance and for providing network access rights to users. In addition, this person should undertake data back-up routines on a regular basis to minimize the effect of data loss due to system failures. These responsibilities should extend to disaster recovery plans and procedures where loss of data jeopardizes the organization's ability to meet its objectives on response time.

## Reviewing and approving data

All data should be examined before use, otherwise you may inadvertently introduce errors into your work. The standard does not require that data controls be the same as

---

[5] *BS 7799 Information Security Management* (British Standards Institution, 1995).

document controls so you are at liberty to pitch the degree of control appropriate to the consequences of failure.

Regarding approval of data, you will need to define which data needs approval before issue as some data may well be used as an input to a document which itself is subject to approval. It all depends on how we interpret 'approved before issue'. Approval before issue should be taken to mean 'issue to someone else'. Therefore, if you alone use data which you have generated then it does not need review and approval prior to use. If you issue data to someone else then it should be dated and reviewed and approved beforehand. If produced using a computer, the file location should also be specified. If your job is to run a computer program in order to check out a product then you might use the data resulting from the test run to adjust the computer or the program. As you should have been authorized to conduct the test in the first place, your approval of the data is not required because the data has not in fact been issued to anyone else. The danger hiding in this requirement is that an eagle-eyed auditor may spot data as being used without any evidence that it has been approved. As a precaution, ensure you have identified in your procedures those types of data that require formal control and that you also know the source of the data you are using.

### Changes to data

As with the review and approval of data, care should be taken with how changes are made. Data that has not been issued to anyone, or has not been used by anyone other than the originator, does not require approval if changed. Only the data that has been issued to, or is used by, someone other than its producer needs to be brought under change control. If you are using data provided by someone else then in principle you can't change it without that person's permission. However, there will be many circumstances where formal change control of data is unnecessary and many where it is vital (as with scientific experiments, research, product testing etc.) One legitimate way to avoid seeking approval to change data is to create new data from old data and assign a new identity to the new data.

## Summary

Document and data control is one of the key elements of any quality system and generates the greatest number of nonconformities during third-party audits. With the advent of computerized solutions to document and data control, the control systems have become complex and sophisticated. Many of the controls are now automated which in some respects removes the burden of work but does not remove the burden of proof to demonstrate that the controls are effective. Electronic documents ease the distribution problems that existed with paper documents but introduce new problems created by the ease with which documents and data can be produced. At one time all work to be typed passed through the typing pool and hence there was some degree of control over what was published. Now, the possibilities exist for anyone to produce a professional-looking document from their desk and, without approval, send it out world-wide, change it and send it out again. Document and data controls have therefore changed in their focus. Gone are the days of scissors and a glue stick, white-out and black ribbons. Some offices and workshops have moved over to a paperless environment, removing the need for visible signatures, change notes, amendment leaflets, revision status, page numbers, issue dates – all this is hidden from the user, who is now reliant on the display being accurate and current because it can't be changed by the user. Even the forms are now completed on-line, the advantage being that the computer won't accept a command to process the form unless all information is entered correctly. There are, however, many offices and factories that have yet to make the transition and others are half and half. Here the old paper controls are still needed. It is ironic that in this computer age, when the prophets told us printed matter would die out, we are using more paper now than at any time since the world began!

### Document control task list

1. Identify the types of document that you need to control.

2. Classify these documents so that you can apply controls appropriate to their classification.

3. Ensure your quality system procedures identify all the types of documents requiring control.

4. Specify appropriate requirements for each of the controlled documents.

5. Establish numbering, dating and revision status conventions.

6. Identify the issuing authorities for the controlled documents.

7. Produce procedures for preparing, reviewing, approving, issuing and changing controlled documents

8   Determine where each type of document is to be stored.

9   Decide how you will indicate the approval status on documents

10  Determine who will review and who will approve the controlled documents.

11  Decide how you will safeguard approved documents against unauthorized change, copying and removal.

12  Create controlled lists of documents which denote the revision status.

13  Create distribution lists for controlled documents.

14  Provide document custodians with stamps to mark obsolete documents upon receipt of instructions.

15  Create a formal change request mechanism for initiating changes to controlled documents.

16  Provide a fast route to change documents.

17  Provide an economic means of changing a range of documents affected by a single change.

18  Provide a means of withdrawing and disposing of documents when the product/organization/service/process becomes obsolete.

19  Provide a means of evaluating the effects that a change in one document has on other documents.

20  Reference the source of any numerical or graphical data, including quotations from books.

21  Treat data sources outside the quality system as external documents.

22  Decide how you will control changes to data that is used in more than one document.

23  Appoint Database Managers and Network Managers where applicable.

## Document control questionnaire

1. How do you control documents and data that relate to the requirements of ISO 9001?

2. How do you control documents of external origin?

3. How do you ensure that documents and data are reviewed for adequacy by authorized prsonnel prior to issue?

4. How do you ensure that documents and data are approved for adequacy by authorized personnel prior to issue?

5. How do you ensure that the pertinent issues of appropriate documents are available at all locations where operations essential to the effective functioning of the quality system are performed?

6. How do you ensure that information on the current revision status of documents is readily available?

7. How do you ensure that invalid and/or obsolete documents are assured against inadvertent use?

8. What means are used to identify obsolete documents retained for legal and/or knowledge preservation purposes?

9. How do you ensure that changes to documents are reviewed by the same functions/organizations that performed the original review?

10. How do you ensure that changes to documents are approved by the same functions/organizations that performed the original approval?

11. How do you ensure that designated organizations have access to pertinent background information upon which to make their review and approval of changes to documents?

12. How do you identify the nature of changes within documents or their attachments?

## Annex A
## Specimen document development procedure

### 1 PURPOSE

The purpose of controlling document development is to maintain the integrity of the management system and ensure that only those documents required for effective management are developed, approved, issued and implemented. This document implements the document development policies defined in section 6.1 of the Policy Manual and responds to clauses 4.5.1 and 4.5.2 of ISO 9001.

### 2 SCOPE AND APPLICABILITY

This procedure covers the initiation, preparation, review, approval, publication and filing of documents that form part of the management system. It applies to control and operating procedures, output documents and standards, forms, labels and notices that form a part thereof. It does not apply to records and registers as these are governed by CP14. It applies to all personnel on all locations and defines their responsibilities and authority, and the methods, forms and channels of communication to be used.

### 3 DEFINITIONS

All general terms and acronyms are defined in ST01.

### 4 PROCEDURE

#### 4.1 Process description

Applications for a new document are passed to the QSM either using a change request or document requirement. Development is controlled through a Development Register in which a number is allocated and an author assigned. The document is prepared to defined standards, comments are collected from potential users, and the document revised and subject to approval by designated authorities. Following approval, the document is registered, translated if necessary, published and filed in designated files under the control of designated custodians. The document development process is illustrated in Figure 1 and indicates the responsibilities and interfaces associated with its execution.

#### 4.2 Identifying the need

The management system is represented by a system model contained in PD02. All system documents shall be derived from this model, either directly or through directly derived documents. The need for a new document would normally arise through a document change request or a system improvement.

Document and data control 267

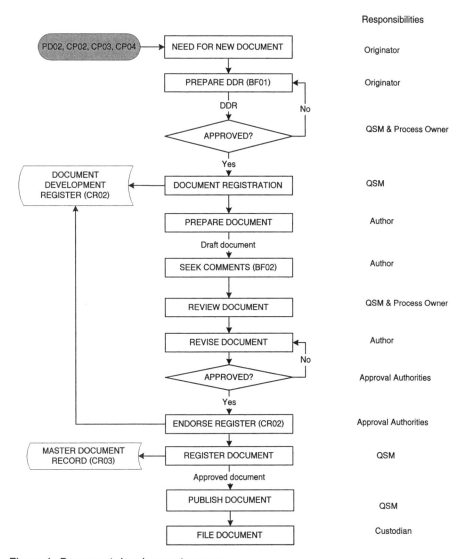

Figure 1 *Document development process*


### 4.3 Document requirement

#### 4.3.1 Application

Following identification of need, a Document Development Requirement (DDR) Form BF01 shall be prepared by the Originator for all new CPs submitted to the Quality System Manager. OPs, ODs, records, registers and forms identified during the development or modification of a Control Procedure do not require further DDRs. CPs apply to all sites and OPs may be site specific but shall relate to the OP invoked in the CP. If a new OP is required that is not derived from the CP then a change to the CP shall be approved before initiating the new OP.

### 4.3.2 DDR approval

On receipt of the DDR the Quality Systems Manager shall assess the requirements and modify them accordingly, add a DDR number and, if justified, complete section 4 after consultation with the process owner. If the need cannot be justified the QSM shall discuss the matter with the Originator. DDRs for CPs require QSM approval. DDRs for local versions of OPs require QSM Representative approval.

### 4.3.3 Document numbering

The document numbering convention to be employed is as follows:

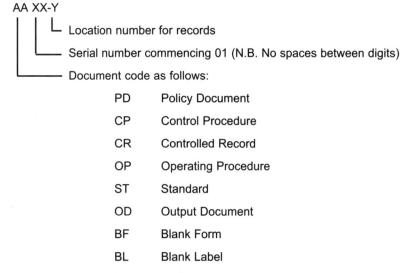

| | |
|---|---|
| PD | Policy Document |
| CP | Control Procedure |
| CR | Controlled Record |
| OP | Operating Procedure |
| ST | Standard |
| OD | Output Document |
| BF | Blank Form |
| BL | Blank Label |
| N  | Notice |

When the purpose of a document changes, a new document number shall be allocated. When the title of a form changes, a new form number shall be allocated. Form numbers given to blank forms may carry a serial number or Output Document number when completed.

Where Output Documents are issued on an annual basis they shall carry the year in the number: e.g. OD29/96, OD29/97.

Where a document is withdrawn, the number can be re-used after the first audit that confirms the change to be fully implemented.

The numbering of contract-specific documents is addressed in OP23.

### 4.3.4 Document development register

Document numbers shall be allocated from the Master Document Register CR02. The author shall be the most competent person available to produce the document. The Process Owners are identified in OD26.

### 4.4 Preparation

The Author may prepare the draft in manuscript or by using a word processor. Manuscript documents shall be converted into MS Word 6 documents prior to being issued for comment. When drafting a document using MS Word 6 an appropriate style sheet shall be selected and text, diagrams and tables entered to meet the requirements of ST10.

### 4.5 Review

The Author shall gather comment on draft documents using Document Comment Form (DCF) BF 02 early in the development process. The DCF shall be sent to members of the System Development Team. On receipt of the completed Comment Forms the Author shall make the necessary changes and circulate a further draft to the Team. Comment Forms shall be retained by the Quality Systems Manager after being actioned and until made obsolete by the document's subsequent revision.

### 4.6 Approval

Following incorporation of changes, the final draft shall be sent to the Approval Authorities indicated below.

| | |
|---|---|
| **Policy Manual** | CEO, MD, QSM |
| **Control Procedures** | MD, QSM, Process Owner |
| **Operating Procedures** | Process Owner, QSM |
| **Standards** | Process Owner, QSM |
| **Output Documents** | As indicated in source procedure |
| **Notices** | Process Owner |

Approval shall be deemed to extend to the forms and labels initiated by the related document.

Where there is no consensus within the Review Panel, the Owner shall decide on the course of action. In the event of the Approval Authorities requiring further changes, the Development cycle shall be repeated until they are satisfied. During this cycle draft issue notation shall be used.

When the Approval Authorities are satisfied a master shall be printed as Issue 1 and a BF15 provided to carry the signatures. OS26 carries specimen signatures of the Approval Authorities. A BF15 is not required where the document carries approval signatures (see ST10). The approval of a System document should indicate that it has not only been found to comply with the prescribed standards but is fit for its purpose.

## 4.7 Registration

Following approval of a document, the status and date of issue shall be entered in the Master Document Register CR02.

## 4.8 Distribution

Following approval of the document, the Quality Systems Manager or his local representative shall obtain the required number of copies and distribute them according to the Distribution List OD01. Additional copies other than replacements shall be marked 'Unmaintained' unless otherwise arranged with the Quality Systems Manager.

## 4.9 Filing

Maintained copies of all System Documents shall be filed in four-ring binders with the following identification:

- Policy Manual
- Procedures Manual
- Standards Manual
- Resources Manual

Each of the above manuals shall contain a Contents List denoting the title and reference number of the documents therein. Each group of documents shall be divided into parts.

The Master paper copies and Master disk files of all system documents first issues (Issue 1), and all subsequent issues, shall be held by the Quality Systems Manager.

# 5 DOCUMENT AND DATA MANAGEMENT

The table below defines the documents used and generated in the execution of this procedure together with the location, retention period and responsibility for their maintenance. The retention period is indefinite unless otherwise stated.

| Ref | Title | Location | Retention | Responsibility |
|---|---|---|---|---|
| BF01 | Document Development Requirement (DDR) | CR04 | 12 months | QSM |
| BF02 | Document Comment Form (DCF) | CR04 | Until next revision | QSM |
| BF15 | Document Approval Record | CR04 | | QSM |
| CP04 | Document Change | PD03 | | QSM |
| CR02 | Master Document Register | CR02 | | QSM |
| OD01 | Distribution List | PD05 | | QSM |
| OD26 | Approval Authorities | PD05 | | QSM |
| OP02 | Management System Data Storage | PD03 | | QSM |
| OP32 | Document Preparation and Change | PD03 | | QSM |
| ST01 | Glossary of Terms and Acronyms | PD04 | | QSM |
| ST10 | Documentation | PD04 | | QSM |

# Annex B
# Specimen document change procedure

## 1  PURPOSE

The purpose of controlling changes to documents is to cause desirable changes and prevent undesirable changes in operational practices.

This document implements the document change policies defined in section 6.6.3 of the Policy Manual and responds to clause 4.5.3 of ISO 9001.

## 2  SCOPE AND APPLICABILITY

This procedure covers the initiation, evaluation, approval and implementation of all changes to internal quality system documentation as well as the withdrawal of such documents. External documents are covered in CP19 and the change procedures for product documentation are addressed within the CP requiring them.

A single request for change may result in a single instruction to change but change requests may also be grouped together to result in a single change notice.

The procedure applies to all staff using quality system documentation.

## 3  DEFINITIONS

All general terms and acronyms are defined in ST01. Those unique to this document are defined as follows:

**Document change**  A change made to the original document.

**Amendment**  Changes made to copies of the original document.

**Change request**  A means of requesting a change to the original document for the purpose of its improvement

**Change notice**  A means of notifying users of documents of changes that have occurred.

**Major change**  A document change the changes the way operations are carried out

**Minor change**  All changes other than major changes

## 4  PROCEDURE

### 4.1  Process description

The document change process is illustrated in Figure 1 and indicates the responsibilities and interfaces associated with its execution.

### 4.2  Identifying the need

A need to change an approved document can arise for a number of reasons:

a) Changes in the business
b) Changes in practice
c) Errors, ambiguity or incompleteness in the original document
d) The impact of changes from other documents

All changes should bring about beneficial improvement in the effectiveness of the management system.

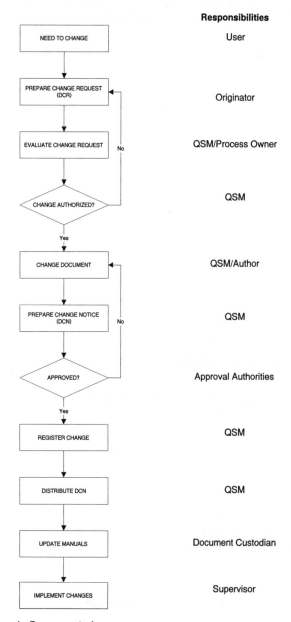

Figure 1 *Document change process*

## 4.3 Change Requests

No change shall be made to the disk, paper master or copy versions of an approved quality system document without being authorized by an approved Document Change Request (DCR). Requests for change shall be made on form BF39 and submitted to the QSM. Changes may be made by providing a marked-up copy of the document or by detailing the changes on the Form.

A DCR may detail the exact nature of a change to an individual document or simply provide a description of a problem that requires a change to one or more documents.

The governing procedure shall apply until the change has been approved or a temporary waiver authorized in writing by the QSM.

## 4.4 Change Evaluation

On receipt of a DCR the QSM shall allocate a serial number, log the DCR on to the MS database, and assign the change to one of the following authorities (depending on the potential impact):

**Level 1** Document owner (No impact beyond the document affected. Covers cosmetic changes and changes that do not affect practices, products or processes.)

**Level 2** Document owner and the functional managers (Impacts practices but confined to the document in which the practices are defined.)

**Level 4** Process Owner and the Product Managers affected (Impacts one process but confined to the documentation which describes the process.)

**Level 5** System Change Board (Impacts more than one process causing redesign of interfacing processes.)

**Level 6** System Review Board (Impacts the system design such that major redesign is necessary.)

In evaluating the change consideration shall be given to:

a) Adequacy of the details given on the DCR

b) The impact on system effectiveness

c) The impact on compliance with the governing requirements (ISO 9002, BS 7750, BS 8800 etc.)

d) The effect on other documents (procedures, output documents, forms and labels affected)

e) The urgency of the change (immediate impact on performance or marginal improvement)

f) The impact on operations at other sites (localized to one site or affecting all sites)

g) Other change requests that have yet to be actioned

h) The method of promulgating the change (hand amendment, page change or document re-issue)

The results of the evaluation shall be entered onto the DCR.

## 4.5 Change authorization

Following evaluation of the DCR the request shall be authorized or rejected by the approval authorities, a copy sent to the originator, and the original filed in File CR34. A copy of all approved DCRs shall also be sent to the process owner.

Should a DCR be rejected, the reason for rejection shall be entered onto the DCR unless the DCR can be made acceptable by alteration in the wording. In such cases the alterations shall be made with the agreement of the originator.

## 4.6 Change document

Following approval of the DCR, the current approved document shall be copied to the Draft Directory and changes made in accordance with section 5.4 of CP01. Changes shall be denoted in the text by using the sidelining feature of MS Word. DCRs may be actioned as and when necessary and the changed document held until sufficient changes warrant a re-issue of the document. Changes that alter practices shall be actioned on a priority basis.

The issue notation on approved documents undergoing change shall be alphanumeric (i.e. Issue 1a, 1b, 1c etc. indicating three revisions of Issue 1).

## 4.7 Prepare change notice

The normal method of promulgating document changes to all sites shall be by re-issue of the changed document under the authority of a Document Change Notice (DCN) Form BF40. The DCN shall be prepared by the Process Owner or the QSM, a serial number allocated and the DCN logged onto the MS database. A DCN may cover changes to several documents within a manual.

Minor Changes affecting small portions of text and not affecting operating practices may be promulgated through an approved DCR pending re-issue of a new version of the affected document.

In the event of it being necessary to promulgate a Major Change in advance of re-issuing the changed documents, such changes shall be specified on a DCN giving the relevant instructions for implementing the change in the document.

DCNs may provide instructions to amend text by hand, to replace pages or to replace a complete document. Hand amendments are acceptable for changes affecting job titles, document numbers, document titles and cross references but should be discouraged for more extensive changes. Page replacement is acceptable for documents over six pages or where the change only affects one page after reprocessing.

Forms and labels should be changed by using a DCN to announce the change.

Should it be necessary to withdraw a document that is not being replaced, the DCN shall relate to the Policy Manual PD02 or to the CP that invokes the document. In such cases, instructions shall be provided to amend all other documents affected by the change.

## 4.8 Change approval

Following preparation of the DCN, it shall be submitted to the same authorities as approved the original document together with the changed document at draft revision status and a copy of the DCRs from which the DCN originated. The approval authorities shall denote approval of the changed document on the DCN. Should further changes be necessary to gain approval, the need for a further draft shall be agreed with the approval authorities, otherwise the agreed changes shall be made and registration process commenced.

## 4.9 Registration of changes

### 4.9.1 Master document register

Following approval of the DCN the revision status of the affected documents shall be added to the Master Document Register CR03.

### 4.9.2 Issue notation

On approval the issue shall increase by 1 (i.e. Issue 1c becomes Issue 2) and the document be copied to the Operational Directory without overwriting the previous versions. The old version shall be moved to the Archive Directory.

Should page replacement be decided, the document issue status shall remain unchanged. The issue status of the pages affected shall be denoted by letter revision (e.g. Issue 1A, 1B). The revision applies to the page, not the whole document.

## 4.10 Distribution

The copies of the DCN and the revised documents shall be distributed according to Distribution List OD01.

## 4.11 Update manual

On receipt of a DCN the document custodian shall implement the instructions and update the Change Record at the front of the appropriate manuals accordingly. The DCN shall then be passed to the Supervisor for action.

Where the change is promulgated through an approved DCR, the DCR shall be filed behind the affected document, the text amended by hand and the DCR number annotated in the margin opposite the change.

## 4.12 Implement changes

On receipt of a DCN from the Document Custodian, the Supervisor shall:

a) Notify staff of the changes.

b) Withdraw any obsolete documents, forms and labels.

c) Install new forms and labels and ensure the new practices are understood.

d) Initiate the new practices and provide appropriate training when necessary.

e) Endorse the DCN indicating the abbreviations have been completed.

f) File DCN in the appropriate manual.

## 5 DOCUMENT AND DATA MANAGEMENT

The table below defines the documents used and generated in the execution of this procedure together with the location, retention period and responsibility for their maintenance. The retention period is indefinite unless otherwise stated.

| Ref | Title | Location | Retention | Responsibility |
|---|---|---|---|---|
| BF39 | Document Change Request (Copy) | Affected Manual | 12 months | Originator |
| BF39 | Document Change Request (Original) | CR34 | 12 months | QSM |
| BF40 | Document Change Notice | CR35 | 12 months | QSM |
| OD01 | Document Distribution | PD03 | | QSM |
| ST01 | Glossary of Terms and Acronyms | PD04 | | QSM |
| CP01 | Document Development | PD03 | | QSM |

## Annex C
## Specimen document change request

| | DOCUMENT CHANGE REQUEST | DCR No: |
|---|---|---|
| Document title: | | Ref: |
| Originator: | | Date: |
| Change requested: | | |
| Reason for change: | | |
| Change evaluation: | | |

| Request rejected | Request approved |
|---|---|
| Signature: | Signature: |
| Date: | Date: |
| Remarks | |

## Annex D
## Specimen document change notice

| | **DOCUMENT CHANGE NOTICE** | **DCN No:** |
|---|---|---|
| **This notice informs recipients that the identified document(s) has/have been changed and the copies in their possession are to be changed as instructed.** ||| 
| **Change Instructions:** <br><br><br><br><br><br><br><br> After incorporating the changes, update the Manual Change Record |||
| **Related DCRs** |||
| **Reason for change:** <br><br><br> |||
| **Change Approval** <br><br> Name . . . . . . . . . . . . Signature . . . . . . . . . . . . . . . . Date: . . . . . . . . . . . . <br><br> Name . . . . . . . . . . . . Signature . . . . . . . . . . . . . . . . Date: . . . . . . . . . . . . <br><br> Name . . . . . . . . . . . . Signature . . . . . . . . . . . . . . . . Date: . . . . . . . . . . . . |||
| Changes implemented and staff trained . . . . . . . . . . . Date . . . . . . . . . . . . <br> <div align="center">Supervisor</div> |||

# Annex E
# Specimen external document control procedure

## 1 PURPOSE

The purpose of external document control is to ensure that only valid documents of external origin are used for determining and implementing the company's policies and practices.

This document implements the external document policies defined in sections 6.4.2 & 6.7.4 of the Policy Manual, clause 4.5.1 of ISO 9001 and 4.7.2 of BS 7750.

## 2 SCOPE AND APPLICABILITY

This procedure covers the acquisition, storage, use, maintenance and withdrawal of external documents. Internal documents are covered by CP01 and CP04.

It applies to all staff who acquire or use external documents.

## 3 DEFINITIONS

All general terms and acronyms are defined in ST01. Those unique to this document are defined as follows:

**External document** A document that is not produced by the company or any of its divisions. It could be a national or international standard, government regulation or document issued by a customer or supplier.

## 4 PROCEDURE

### 4.1 Process description

The external document control process is illustrated in Figure 1 and indicates the responsibilities and interfaces associated with its execution.

The principle means of controlling external documents is through a number of registers. Four classes of external documents have been identified. The only external documents that are referenced in the management system documents are standards and regulations and these are identified in one register under the control of the QSM. Customer documents, software manuals and equipment manuals each have separate registers under the control of the relevant process owners. Measures are taken to evaluate all new and revised external documents and where necessary activate improvement programmes to implement new or revised standards and regulations. Copies are strictly controlled and provision made for retaining and disposing of obsolete documents.

### 4.2 Identification

External documents that apply to the company's operations are defined in PDs, CPs, OPs, STs or ODs. In general no other documents should be needed. Should reference need to be made to an external document on a regular basis that is not listed in any of these documents, then a DCR (BF39) shall be initiated to reference the document in

**280** Document and data control

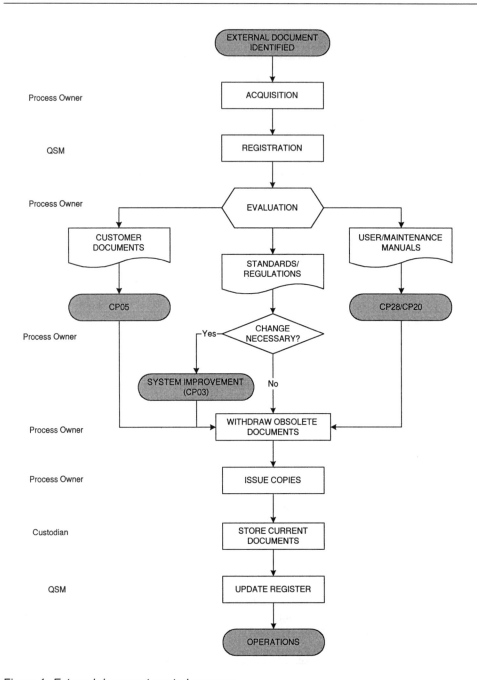

Figure 1 *External document control process*

the relevant company document and add it to the External Document Register CR46 (see below).

## 4.3 Acquisition

External documents can be acquired through a number of different routes:

| | |
|---|---|
| **Purchase** | These shall be obtained through Purchase Order in accordance with CP07. |
| **Customer** | These will be issued by the customer with the contract or may have to be obtained on request from the customer. |
| **Supplier** | These may be supplied with equipment purchased by the company for use in using and maintaining such equipment or may have to be obtained on request from the supplier. |
| **Free issue** | These may be received from government agencies or may need to be requested from such agencies. |

## 4.4 Registration

All external documents and amendment thereto shall be registered in an External Documents Register (CR46). Four such registers shall be established and maintained:

- Customer Documents Register (CR46-1) – maintained by Sales
- National and International Standards and Regulations (CR46-2) – maintained by the QSM
- Plant and Equipment User and Maintenance Manuals (CR46-3) – maintained by the Maintenance Manager
- Software User Manuals (CR46-4) – maintained by the IT Manager

The Register shall contain the following data as appropriate:

- Document Reference
- Title
- Revision Status & Date
- Issuing Authority
- Identity of documents in which it is referenced
- Custodian
- Distribution of copies
- Effectivity (current, obsolete or a stated future date)

## 4.5 Evaluation

### 4.5.1 Customer documents

When a new or a new version of a customer document is received, it shall be reviewed and dispositioned in accordance with CP05.

### 4.5.2 Plant user/maintenance manuals

In the unlikely event of a plant/equipment supplier issuing a revision of a user/maintenance document it shall be reviewed and dispositioned in accordance with CP28.

### 4.5.3 Software manuals

Should any amended data in software user manuals be received, it shall be reviewed and dispositioned in accordance with CP20.

### 4.5.4 Standards and regulations

When revisions of national and international standards and government regulations are received they shall be reviewed by the QSM and the nature of change and its effectivity date determined. If the change has no material effect on policies and practices the new version shall supersede the obsolete version. If the changes are perceived as potentially affecting policies or practice and there is no date by which the provisions become mandatory on the company then its adoption shall be at the discretion of the QSM.

If the changes are perceived as potentially affecting policies or practice and there is a date by which the provisions become mandatory on the company then the QSM shall initiate an Improvement Plan in accordance with CP03.

## 4.6 Obsolete documents

### 4.6.1 Use of obsolete documents

Following determination of the action required on receipt of a new or revised external document, the Process Owner shall determine whether obsolete versions are to be retained for knowledge purposes. Those to be retained shall be identified in the appropriate Register and in order to prevent their use they shall be held by the Process Owner only.

### 4.6.2 Obsolete equipment/software manuals

When plant, equipment or software is replaced and discarded the associated manuals shall also be discarded.

### 4.6.3 Obsolete standards and regulations

A new or revised document does not make the old one obsolete until any new policies and practices resulting from the changes are implemented. It is therefore likely that both old and new versions will be in use concurrently for a transitional period.

The DCN (BF40) issued, indicating the changes to be made to existing policies and practices, shall indicate the action to be taken on the previous versions of the external documents and distributed copies if any. Such action shall also be indicated in the appropriate Register. The Registers shall be the indicator of currency and obsolescence regarding external documents.

## 4.7 Issuing copies

Photocopying of external documents shall be limited to that permitted by the copyright notice displayed in the document concerned.

Software user manuals shall be issued to PC users appropriate to the software installed.

The Register shall indicate the distribution of controlled copies in use.

## 4.8 Storage

National and international standards shall be stored under the control of the QSM in labelled files, where appropriate.

Government regulations shall be stored under the control of the QSM in labelled files, where appropriate.

Contract documents shall be stored under the control of Sales in the relevant customer file.

Equipment manuals shall be stored under the control of the Maintenance Manager (MM).

Software manuals shall be stored under the control of the IT Manager (ITM).

## 4.9 Use

External documents shall be used to the extent indicated in the appropriate company document. Some documents are not used in their entirety and provisions may not apply to the company's operations.

## 5 DOCUMENT AND DATA MANAGEMENT

The table below defines the documents used and generated in the execution of this procedure together with the location, retention period and responsibility for their maintenance. The retention period is indefinite unless otherwise stated.

| Ref | Title | Location | Retention | Responsibility |
|---|---|---|---|---|
| BF39 | Document change request | CR05 | 12 months | QSM |
| BF40 | Document change notice | CR05 | 12 months | QSM |
| CP01 | Document development | PD02 | | QSM |
| CP03 | System improvement | PD02 | | QSM |
| CP04 | Document change | PD02 | | QSM |
| CP05 | Acquisition | PD02 | | SM |
| CP07 | Purchasing | PD02 | | PM |
| CP20 | Electronic data control | PD02 | | ITM |
| CP28 | Plant maintenance | PD02 | | MM |
| CR46 | External documents register | CR46 | | QSM |

## Chapter 8

# The quality system management process

Previous chapters of this book have covered quality system design and documentation and the preparatory and project management activities associated with developing a system using the process management approach. As each organization is different, individual business processes have not been covered in any detail, the flowcharts being provided merely to serve as examples. Quality system management, however, is likely to be a new subsystem in an organization that does not have an established quality system and is therefore treated in detail in this chapter.

### The process model

Quality system management consists of a range of processes that serve to establish, implement, maintain and improve the quality system. A generic subsystem model is illustrated in Figure 8.1. This is a modified version of the diagram provided in the *ISO 9000 Quality Systems Handbook*.

The system development process was addressed in Chapter 5. The system implementation, evaluation, maintenance and improvement process will be covered in this chapter.

### System implementation

#### The implementation process

The System Implementation Process is illustrated in Figure 8.2. The grey boxes indicate interfaces with other processes. Following development of a process, it is prepared for implementation. The documentation is then released to users through a commissioning process. There are two phases of implementation that are not shown on the diagram. There is *initial implementation* (which precedes a qualification and hand-over stage) and

# 286  The quality system management process

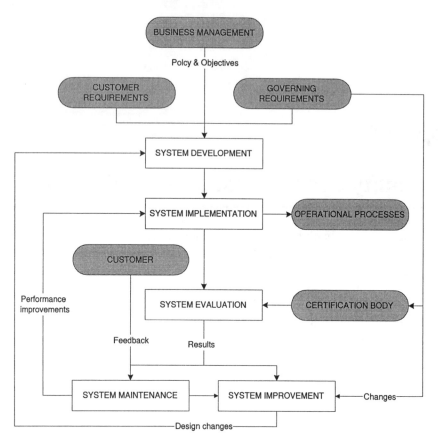

Figure 8.1  *Quality system management process*

*final implementation* (when the system is stable). The introduction of any changes arising from the Document Change Process also need to be planned – hence the separate steps before implementation, although only major changes will need such formality. The data generated by the system feeds the System Evaluation Process and any problems in turn are fed into the System Maintenance Process for resolution. External documents enter the System Implementation Process through the External Document Control Process. Documents required by implementing the practices enter the Document Development Process and any records required enter the Records Control Process before being used by the System Evaluation Process. This may appear rather complicated but it serves to organize information that is produced when implementing the system. The philosophy throughout this book has been to adopt the 'process management approach' to quality system management and therefore each process has to link with other processes. The inputs enter a process from other server processes with their outputs feeding receiving processes. If inputs and outputs have no receiver then the process won't function effectively. It will rely on informal actions to collect and route information and, hence, will not produce a robust system.

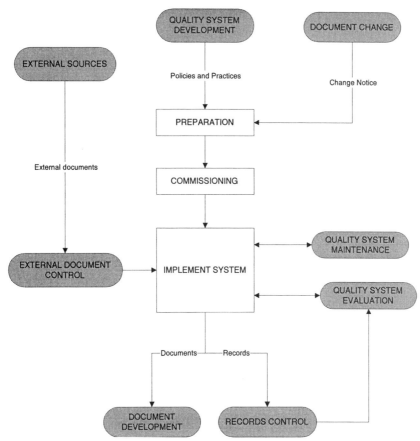

Figure 8.2 *Quality system implementation process*

## Implementation strategies

Implementation follows documentation development. It should be a phased implementation commencing when the first documents are ready to implement and completing when the system is fully in place and all aspects are operational. Each process should undergo a phased implementation that commences with preparation, followed by a commissioning phase where the systems are put in place. This then leads to a period of implementation where practices are monitored. At a suitable stage, the practices should be subject to qualification as proof that they are effective. They can be subsequently handed over to the line managers for continuous implementation and improvement. This is one strategy that is used for launching any new systems into operation, be they computer systems, communication systems, transportation systems or any systems that have received a major investment. The same approach is also used with new products and services as developers desire a smooth introduction into service. Send a product to the market before its development is complete and the subsequent customer feedback will indicate you were too hasty! The rush to get products on sale is often counterproductive. The same can be true with quality systems where there is a rush to get procedures issued before they have been proven. In some cases several months are spent fine-tuning the practices and the system abounds with informal methods and transactions. There are several advantages in applying the design control principles of element 4.4 of ISO 9001 as they include some fundamental concepts that, if used diligently, will inject the right degree of control over quality system development. The requirements of clause 4.4 that can be applied to system implementation are those concerned with design verification and validation.

Many companies simply release the documentation and then start auditing only to find that people were not prepared for the new practices and resent being measured against criteria about which they were not advised in advance. A typical result might be the manager who says that the procedures cannot be implemented because no provision was made in the budget to acquire the necessary equipment or recruit the necessary personnel. For the most part, however, the procedures will reflect what people already do but there will be occasions where a lack of preparation will result in delays. The system should not be launched as a big bang but more as a trickle as elements of the system are put in place.

## Preparation

Implementation is the process of applying the documentation to the processes and tasks that are to be performed. With some processes the documentation may well describe what already takes place and so no preparation is necessary. Where new practices and processes are to be introduced, care needs to be taken in introducing these to the work place and careful planning will be needed.

Each document has an applicability and hence the functions concerned need to be informed and any resistance to change overcome before launch to ensure compliance.

The first practices to implement will be the document development and control procedures followed by those associated with the first business processes. When planning implementation, you need to determine the inputs of a process, ensuring these inputs are available from the interfacing process, otherwise the staff cannot implement the procedure as stated. Closely behind the document control practices should be the auditing practices so that auditing can be used to qualify the processes.

If you have designed databases to contain the documentation and capture records and other data then obviously these need to be launched before the documentation is released. As mentioned in Chapter 4, a Problem Management tool should be used to collect problems throughout development so that solutions may be conveyed to staff in a uniform manner. During implementation this tool will become most valuable as problems emerge, providing staff with a means to report problems and obtain resolutions. The same Problem Management tool can be used to capture customer complaints, nonconformity data, supplier problems etc. The sooner you start entering this data into a database the better since it will serve to demonstrate improvements as a result of introducing a formal quality system.

The following are some of the activities to be carried out when preparing for the introduction of new or revised practices:

- Establish who is affected by the procedure and verify that they know what is expected of them.

- Identify the areas where resistance is likely and develop the appropriate responses.

- Choose the timing of the change and identify the champions for each department.

- Decide whether to conduct trial installation in one area, progressive installation across all areas or a big bang approach, and identify who is to be involved.

- Determine the extent of staff training required.

- Consider the preparation of staff awareness bulletins (advertising).

- Advise managers on the approach they should take in announcing the changes to their staff and making them aware of their new responsibilities.

- If a major change affects many areas (or significantly affects one particular area) the changes should be added to the Annual Quality Programme and detailed plans drawn up on how the old practices will be gradually changed and/or new practices introduced.

- Determine what other procedures have to be in place for the new procedure to work effectively.

- Determine how commitment is going to be tested.

- Decide what communication links are necessary for reporting problems, dealing with priorities and disputes, and processing changes.

- Decide when the old practices are to be discontinued. (It may be wise to leave the old system in place until the transition is complete.)

Always conduct a dry run to test the procedure. This is especially true with electronic systems as database changes are more difficult to implement after data has been placed in the database. Make sure you have all the resources available to implement a procedure as, again, it won't work if the tools, people etc. are not in place. The plan therefore needs to make provision for purchasing equipment, recruiting or training people etc.

## Commissioning

Commissioning is the process of setting the processes to work. With electronic systems the release process will put the documentation out to everyone so you may not have a choice. The old documents need to be withdrawn and new ones put in place progressively. Most importantly, ensure ownership rests with the users, so persuade them to make the changes in practice.

Either during or after the preparation process and, depending on the type of procedure to be introduced, several activities may be necessary before full implementation is possible:

- New forms should be introduced and the old ones removed from circulation.

- New files should be established for holding the new data and existing files should be re-identified.

- The communication links should be set up and verified as those supplying or receiving information associated with the procedure may be unaware of their new responsibilities.

- New registers of records should be created if required.

- New distribution lists may be required.

- New equipment, tools and facilities should be acquired and commissioned.

- The awareness briefings prepared previously should be put into effect.

- The training plans prepared previously should be put into effect.

- Commissioning is finished when everything is in place ready for implementation.

## Initial implementation

Following the Commissioning Phase everything should be in place to effect a smooth implementation. Once the processes are in place you need to monitor them closely, observe how people use them and then resolve any problems. Awareness sessions need to continue so that everyone affected is covered. As implementation will be progressive, the interfaces between processes, functions and sites becomes a critical issue. Don't just issue a procedure and sit back. Nothing will happen! You need to check that the people understand what is required and it may take some time to confirm that the new procedure really is being used.

Provide a feedback mechanism so the users can convey their concerns and get them resolved quickly (use the Problem Management tool referred to previously). In a large organization, this task will need more than one person. Sending out messages will not work; you need to go to the people and talk with them and hence you will need support from their supervisors first. Get supervisors on board to do the awareness training.

During this phase the actions and decisions required by the procedure are carried out and this may initially demand considerable effort as the following may need to be addressed:

- New documents may have to be produced and old ones may have to be revised or reformatted. This may mean a major upheaval until all documents have been completed and the process stabilized.

- Staff performing new activities for the first time will reveal problems with the procedure or the training which will need to be resolved.

- Difficulties may emerge at the interfaces.

- The procedure or parts of it may not be activated for some time.

During this phase the activities should be closely monitored to ensure that the new practices remain in place and staff do not revert to the old practices. There will be a tendency for people to return to their old practices unless the managers constantly monitor the implementation of the new practices. You cannot rely on auditing. Auditing discovers errors after the event and it may be months after the event. Where staff have been using the old practices, a change in direction can set you back weeks whilst you conduct retraining. Staff may not have adopted the new practices well and without monitoring may develop bad habits. Care should be taken to ensure that the intent of the procedure is understood and that any resistance to change is overcome.

## Qualification

The qualification stage commences when a process is in place. It is the stage where tests are carried out to verify that the process is performing as expected. Process qualification is an element of Quality System Evaluation. The qualification stage is a term used to distinguish the initial proving audits and reviews from the routine audits and reviews. The new procedures should be qualified or validated; that is, they should be thoroughly tested to verify that they fulfil their purpose. Auditing should commence following implementation. You can commence audits beforehand, but you will not be testing a stable system. Asking questions of people who are without awareness of the requirements to be met only builds up resistance unless you explain that the audit is being used as a tool to find the gaps and problems within the system. Use the formal auditing system so that this is qualified at the same time (see *System audit* later in this chapter). At this stage, the audits should be against the appropriate ISO 9000 standard as well as the approved policies and practices so as to determine the extent of compliance before the third-party audit.

When sufficient audit data has been generated, the first System Review can take place. Before this time, there is no data on which to judge the effectiveness of the system, unless you already have a customer feedback and performance assessment process in place.

The implementation of every requirement of the procedure should be verified in every area to which it applies to determine the following:

- Which areas are applying the procedure.

- The extent to which the procedure is being followed.

- The extent of variation between different departments, if more than one department implements the procedure.

- Whether the procedure if followed as stated will produce the required results, consistently and predictably.

- Whether the required results are still valid in light of changing circumstances.

- Which changes are needed to make it effective.

If as a result of the above no changes are necessary either in the procedure or in the practice then the procedure is qualified for its application. If, at some later stage, the procedure is applied in new areas, a similar review should be undertaken after commissioning. If changes are necessary, procedure implementation should be re-examined after the changes have been implemented.

**Hand-over**

Not until all documented practices are being implemented and audited can you consider that the process is in place. The initial problems discovered during the audits should have been resolved and the changes made and verified as correcting the problem. A new audit schedule needs to be created to reflect the fact that the process is now stable. Most of all, give encouragement to those working the new system. Use them as examples in cases where you still have resistance. Using the implementation plan as a record, you can indicate which processes have completed implementation.

When the practices are handed over to the line managers they assume control and should maintain the system as described in the section of this chapter dealing with system maintenance. When all the practices have been successfully qualified and handed over, the project team can be disbanded as they have done their job. The management of the system thereafter becomes the responsibility of the Quality System Manager and his/her team (also discussed later in this chapter).

## System implementation task list

1. Prepare implementation plan.
2. Implement by process not function.
3. Identify resistance to change.
4. Plan progressive documentation release.
5. Acquire necessary resources to implement.
6. Conduct dry run.
7. Maintain action list.
8. Ensure understanding.
9. Monitor execution.
10. Resolve problems.
11. Continue awareness and training sessions.
12. Check effect at the functional interfaces.
13. Start in areas of least resistance.
14. Set up computers/equipment.
15. Withdraw obsolete documents and data.
16. Organize paper and electronic data.
17. Commence conversion, where necessary.
18. Ensure ownership rests with users.
19. Commence audit program.
20. Audit frequently.
21. Analyse results.
22. Gather feedback from users.
23. Determine process and system effectiveness.
24. Take remedial and corrective action.
25. Set audit frequency for post qualification.
26. Update training records.
27. Update document status records.
28. Give encouragement.
29. Move on to next process.

## System implementation questionnaire

1 Have you identified which functions will be affected by the changes?

2 Have you briefed the managers of the changes?

3 Have you trained staff in the use of the databases?

4 Have you notified staff what to expect from the internal audits?

5 Have you created prototype databases on which to prove any changes found during implementation?

6 Have you communicated with all affected sites?

7 Have you got the commitment of the managers?

8 Have you identified what changes will be made to existing practices and communicated this to the relevant staff?

9 Have you trained the auditors and analysed their results before extending the audit programme?

10 Have you installed the problem management database or reporting procedures?

11 Have you initiated document control?

12 Does everyone affected know what to do?

13 Are you monitoring progress against plan?

14 Are you raising concerns with the management and getting prompt action?

15 Have you identified the areas of resistance and are you dealing with them before they jeopardize your programme?

16 Have you removed the obsolete documents from the offices and the file servers?

17 Have you decided how you will measure system effectiveness and put the data collection channels in place?

## System evaluation

### The evaluation process

The evaluation process is illustrated in Figure 8.3. The evaluation process follows Quality System Implementation and, as stated previously, should be in two phases: the initial evaluation (called *qualification*) and the routine evaluation. The process consists of four separate processes. The System Audit Process determines whether the policies and practices are being followed and feeds results to the System Review Process for evaluation and to the Quality System Maintenance Process for action in parallel. The External Audit Process is concerned with the activities of the appointed Certification Body or Registrar. This also feeds data into the System Review Process for evaluation and to the Quality System Maintenance Process for action in parallel. The Performance Assessment Process takes performance data from the Operational Processes and determines if the quality objectives have been met, feeding the results into the System Review Process for evaluation and into the Business Management Process in parallel for assessment against the business goals. The System Review Process itself equates to the Management Review of ISO 9001 and takes all these inputs to determine the effectiveness of the quality system. The results of the system review are fed into the Quality System Improvement Process in parallel with other inputs from System Maintenance Process as shown in Figure 8.1.

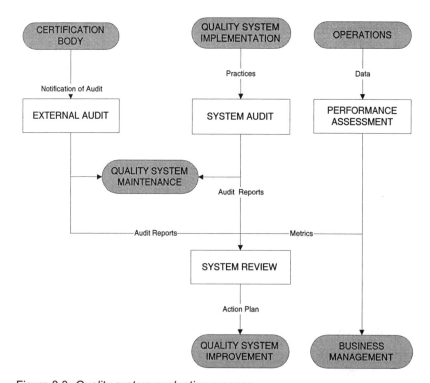

Figure 8.3 *Quality system evaluation process*

## Evaluation strategies

The primary vehicle for system evaluation is the audit process, both internal and external. Many organizations do not include an External Audit Procedure in their quality system because they merely react to the results and pass them through the established document change process. However, the external audit does form part of your system evaluation process and should not be ignored simply because it is external. There may well be other external audits that can input data such as product audits and financial audits. Data arising from these should also be evaluated for opportunities for improvement in the quality system.

ISO 9001 separates the system evaluation elements of the system and, in so doing, causes some confusion (clauses 4.1.3 & 4.17). The management review is required to address system effectiveness but so is the internal audit element. The management review is also required to establish that the system satisfies the quality policy and objectives but not the internal quality audits. However the definition of quality audits in ISO 8402 does state that they should verify that *the arrangements are suitable to achieve objectives* but there is no mention of any measurements required to demonstrate that the objectives are being achieved – hence the necessity for including a Performance Assessment Process. Many systems feed nonconformance data and customer complaints data into the management review with the data collection process being part of the review process. This will work but as the management review is often conducted annually, the process would be exercised only once each year. With a separate Performance Assessment Process, you can assess performance more frequently and feed the improvement process whenever a situation requires action.

## System audit

### *System audit strategy*
All audits are performed to establish facts rather than faults. They aim to establish, by an unbiased means, factual information on some aspect of performance. As the performance of any organization will degrade unless some checks are carried out, audits are performed as a safeguard against a deterioration in standards. Hence, all audits will detect variation from predefined standards. Some variation may be tolerable, other variation may be unacceptable and, if left to continue, may well signify loss of business, credibility and customer confidence. For audits to be useful they have to be against standards which the organization is committed to meeting, otherwise they will not be taken seriously. Internal quality system audits should be performed against the approved policies and practices that comprise the documented quality system. They may also be conducted against the appropriate ISO 9000 standard but this should not be necessary except during system qualification (see *Quality system maintenance*). The purpose of the system audit is to determine the extent to which these policies and practices are being followed. Audits provide data that can be used to determine the quality system's effec-

tiveness in implementing the quality policy, thereby enabling the organization to meet its quality objectives[1].

The standard requires internal quality audits to be planned and implemented and ISO 8402 defines what a quality audit is. Although the definition relates to audits of the quality system, the notes that follow the definition imply that process audits, product audits and service audits fall within the scope of the definition. However, product audits are more like independent inspections than audits against practices and in ISO 9001 the intention is that the system be audited. Although one could interpret the statement in clause 4.17 as requiring audits of products and processes as well as systems, in practice the internal audit programme is usually confined to audits of the system. The difference between a system audit and a product audit is that the system audit establishes that provisions are in place to cause conformity and prevent nonconformity whereas the product audit verifies that product requirements are being met. In the system audit, there is little, if any, examination of products against specification. In the system audit, confidence in product quality is obtained through:

- Evidence that the system caused the generation of a specification.

- Evidence that the specification has been approved by authorized and qualified personnel.

- Evidence that inspections were carried out by trained personnel to approved documented procedures.

- Evidence that the documented results indicate conformity.

Only a sample check of parameters may be made to verify that the right decisions were made. In a product audit, a sample product is examined for compliance with all the requirements of the specification. Hence the use of the term *system audit* so as to avoid any ambiguity.

### System audit process
A typical system audit process is illustrated in Figure 8.4.

### Audit schedule
An audit schedule should be prepared so as to plan the audits to be conducted throughout the year. ISO 9000 requires audits to be scheduled on the basis of status and importance[2]. This means that the frequency of audits should be dictated by the maturity of a process, procedure or activity, its performance with respect to previous audits and how important it is relative to the business. Initially your audit schedule can be devel-

---

[1] For further details, see *ISO 9000 Quality Systems Handbook*.
[2] See *ISO 9000 Quality Systems Handbook*, Part 2 Chapter 17.

*The quality system management process* 299

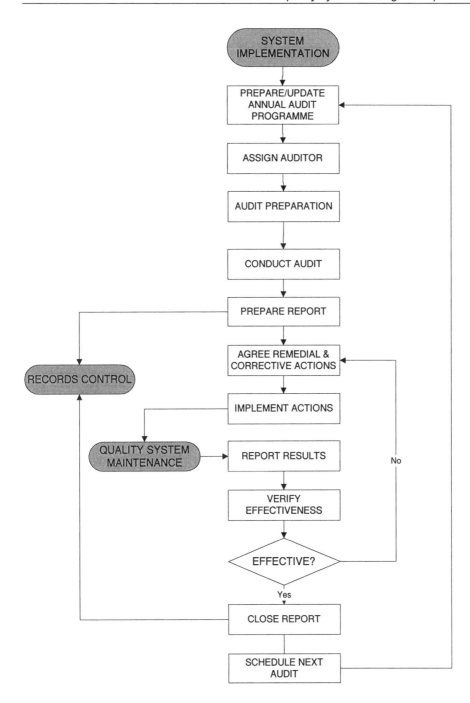

Figure 8.4 *System audit process*

300  *The quality system management process*

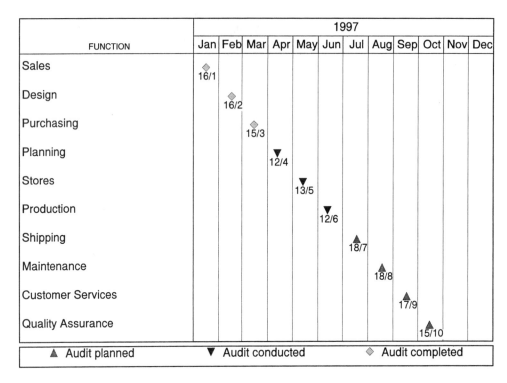

Figure 8.5  *System audit schedule*

oped using a set audit frequency, such as once per month. But as you gather the results, you should modify the frequency to reflect status and importance. An example of an audit schedule is given in Figure 8.5.

The audits may be carried out on a functional (departmental) basis, by process or by procedure. Some internal audits are planned by the elements of ISO 9001, 4.1 to 4.20. This is how many of the certification bodies carry our their audits, although one cannot audit strictly by element since many elements affect more than one department and to apply this approach would mean that you would check 4.1, 4.5, 4.16 and 4.18 in every department. Therefore, scheduling a month when you would check element 4.5 would require you to audit every single department. It is possible but you need to allow sufficient time to cover the relevant elements in detail. A better way is to audit by process as it causes you to cross functional boundaries and check the interfaces. Process-oriented audits are more difficult if the processes cross site boundaries as well as department boundaries. The logistics create a timing problem not to mention the cost if long distances are involved. It would be impractical to audit by process if you had to fly between locations as is the case with some American corporations.

The advent of *groupware* products such as Lotus Notes enables organizations to communicate in real time and it matters not where the people are located. Some may work from their own home and others from hotel rooms. It therefore makes no sense to insist on visiting each location where people carry out their work when the location is irrelevant to the results.

When auditing by function you can still check the processes and schedule the audits so that you follow a trail through the functions from customer input to product/service delivery. The functions on the schedule need to be the same as identified in the Functional Matrix during system development. If they are different then the matrix needs to be updated.

You may need more than one schedule if you target projects, contracts, products and processes in your audit programme.

You should have produced a system audit policy for your Policy Manual. This should indicate the intentions of the audit programme in terms of its coverage over a given period. You may, for instance, have stated that the audits will examine every policy and procedure once each year wherever it applies. This would be truly comprehensive but also very ambitious in a large company. The schedule needs to reflect your audit policy and if you don't have sufficient number of auditors you could be in breach of your policy. The audits will only sample operations but you are not auditing the system if you don't cover all the policies and procedures over a given period. Hence, the results of the audits should indicate the coverage so that you can determine how comprehensive the audit programme is and adjust it to comply with your policy.

When scheduling audits for the annual plan, you need not send out notices to the managers as you need not fix the exact date at that time. The date should be agreed between the auditor and the manager later.

### Assigning the auditor
The audits should be assigned by the planner (the Quality Manager or Audit Manager) on the basis of knowledge of the function, project, product or process. For ISO 9000 element-based auditing you will probably need a team of auditors and, for complex areas, you will need specialists to support the auditor. The auditors need to be trained but do not necessarily need to be qualified lead auditors registered with the national auditor registrar. The training needed is up to you but many companies send their nominated auditors on external training courses to acquire the skills[3]. With novice auditors (auditors straight out of training) you need to send them out with an experienced auditor so that they can learn on the job. (No training course yet offered includes enough practical work and experience to gain competency.)

---

[3] For details of auditor training, see *ISO 9000 Quality System Assessment Handbook*, David Hoyle (Butterworth-Heinemann, 1996), Chapter 2.

Depending on the size of your organization, you may need several auditors. They can be full-time employees, contractors or part-time employees. If part-time, a good approach is to train one or more auditors from each department. In this way you can maintain independence by using auditors from different departments to audit the audit management system. If you only have one auditor then who audits the auditor? You can't rely on the Certification Body but you can use a contractor.

## *Audit preparation*

Once assigned, the auditor should prepare for the audit, reviewing the relevant documentation to identify what needs to be checked. It is often insufficient to simply select the departmental procedures. Many policies and procedures may apply to more than one department and a good guide will be the Functional Matrix prepared during system development. The auditor should prepare check lists to show the coverage of the audit. These are useful later when checking what was audited. A file of nonconformance reports only tells you what was found to be nonconforming, not what was checked and found compliant. ISO 9001 requires the results of the audit to be reported and this means the conformities as well as the nonconformities. Before conducting the audit the auditor needs to inform the managers concerned to agree a date. It is advisable at this stage to inform the manager not simply that an audit will take place in his/her department but what the objective is and what policies and procedures will be checked. It is also important to indicate the time that will be needed to gather the objective evidence. If you tell the manager 'It will take only one hour' ensure you do only take an hour! Be careful to estimate correctly. If you haven't gathered sufficient evidence in one hour then request more time. If more time is not available, then either reschedule the audit or adjust your objective accordingly.

There is a tendency for inexperienced auditors to spend a short amount of time and experienced auditors to spend longer but not proportionately so. The inexperienced auditor doesn't probe deeply because they are unsure what they are looking for. The experienced auditor probes deeply but much more quickly because he/she knows what to look for.

## *Audit conduct*

Before carrying out any audits the managers concerned should have been informed of what to expect. This is usually conveyed at an Opening Meeting. With internal audits, these meetings need not be very formal but you must ensure that the manager is aware of what will be reported and to whom it will be reported. Following the audit, the results should be reported to the manager to obtain commitment to any actions that are required. Unlike external audits, there is no recommendation for certification but a verbal agreement on the findings is necessary.

Train your auditors to apply common sense to their findings. There is nothing that irritates managers more than being drawn into arguments over trivia. Make sure you have found something that really needs correcting, not a random failure that can be corrected on the spot. Good auditors will often point out random failures to the auditee for them to correct immediately and will not issue a report. Remember you are trying to establish the extent of compliance so take many samples. One sample will not give you sufficient objective evidence on which to base a conclusion[4].

### Audit reporting

The most common means of reporting audit results is to use an Audit Findings Report that defines the problem, requests action and provides for follow-up and closure. An example of such a form is illustrated in Figure 8.6.

When documenting your findings you need to ensure that you have provided sufficient details on which the manager can take action. Vagueness only irritates managers since they will not understand what action to take. Be specific as to the incident, the unsatisfied requirement and the location (where necessary). The incident should contain as much information as necessary for the manager to locate the problem you found. You will notice that the report in the example is called an Audit Findings Report rather than a Nonconformity Report. If you were to use terms like 'nonconformance' you may well cause adverse reaction with the auditee. As the form is used to report findings that are not nonconformities (that is, observations) then a label of 'Nonconformity Report' would be misleading.

### Classifying findings

The external auditors often classify their findings as Observations, Minor Nonconformities and Major Nonconformities. You do not need to do the same. You do not, in fact, need a classification convention at all but it is useful to prioritize the findings. Some Certification Bodies base their classification on ISO 9000 such that a major nonconformity is the absence or total breakdown of a system to meet the requirements of a clause of ISO 9001. If you are not auditing against ISO 9001 or ISO 9002 then this classification would not be appropriate. You should have a system that makes sense in your organization. One solution is to classify by effect. For example, a major nonconformity is a departure from requirements that will directly affect the quality of the service or product supplied. In this way you can demonstrate to management the effectiveness of the audit programme in finding problems before they affect business performance.

### Conclusions

Each audit should result in a conclusion otherwise they may be perceived as a witch hunt. From the objective evidence you have found you should be able to draw some conclusions about the extent of compliance in the function or process you audited. If you have only taken one sample, then this will be inconclusive unless there was only one example to audit.

---

[4] See *ISO 9000 Quality System Assessment Handbook*.

| | | **AUDIT FINDINGS** | Audit Ref |
|---|---|---|---|
| | | | Page    of |
| Ref : | Location: | | Auditee: |

| |
|---|
| **Findings** (include subject, incident, requirement, requirement reference and precise location if necessary) |
| |
| **Classification**    Major nonconformity    Minor nonconformity    Observation<br>☐                                    ☐                                ☐ |
| **Acceptance** |

| *Auditor* | *Auditee/Manager* |
|---|---|
| Signature | Signature |
| Name | Name |

| |
|---|
| **Root cause** |
| **Proposed action** (including the specific remedy, the action on others in the population and action to prevent recurrence with implementation dates) |
| Supplementary sheets |
| **Follow-up action** |
| **Proposed action accepted**   Date: . . . . . . . . By: . . . . . . . .   Signature: . . . . .<br>**Proposed action taken**         Date: . . . . . . . . By: . . . . . . . .   Signature: . . . . .<br>**Proposed action verified**     Date: . . . . . . . . By: . . . . . . . .   Signature: . . . . . |

Figure 8.6 *Audit Findings Report*

## Root cause

When deficiencies have been found the managers should propose action to eliminate them but not before establishing the root cause. Many audit records do not make provision for recording the root cause and this can lead to the corrective actions not being completely effective. Anyone can perform the analysis to find the cause of the problem

## Remedial and corrective action

In ISO 9000 the action taken to eliminate the cause of the nonconformity is called *corrective action* but the definition of corrective action in ISO 8402 limits the action to that needed to prevent a recurrence of the problem. Managers therefore need to take four actions as appropriate:

- Primary remedial action (action to correct the nonconformity found)

- Secondary remedial action (a search for other similar occurrences and their correction)

- Temporary corrective action (action to prevent immediate recurrence such as posting notices)

- Permanent corrective action (action to remove the cause and thus prevent recurrence)

The auditor should check that these actions have been specified before accepting the proposals and that the root cause is valid. The implementation of these actions will follow the Quality System Maintenance Process (see *Quality system maintenance* later in this chapter)

## Audit follow-up

After the due date for completion of the actions the auditor should check they have been carried out. With an on-line system, this may be possible by searching the database; otherwise the auditor may need to contact the manager to seek progress. When action completion has been confirmed, the auditor should carry out a follow-up audit to verify that the actions have been effective in eliminating the problem. This can be carried out either at the next planned audit or at an earlier date if the problem found is serious.

## Auditing tools

There are several electronic auditing tools on the market designed around proprietary database software. They enable you to plan the audit, schedule individual audits, produce check lists, record and report findings and proposed corrective actions, and monitor implementation. Many are set up to operate using an e-mail system so that information is routed to individuals on a particular network. The main problem with these tools is that the degree of customization is limited. If you procure different tools for different tasks, you may find they each have communication and problem management features that are not mutually compatible. Your optimum solution may be to use one tool for these functions so as to enable you to share data. Some packages include analysis tools that enable you to sort by any number of categories and produce pie charts, histograms etc. If you intend to design your own tools or select from those available then here are some tips:

- Ensure you can plan your audit by process, function, project, contract, product or any aspect of the business. Beware of those tools that limit you to planning by element of ISO 9001!

- Ensure you can enter your audit findings in free form text with options for entering procedure references, clause numbers of ISO 9001 and key words. You need to be able to sort findings on a number of fields not just elements of ISO 9001 and functions.

- Ensure you can customize the action plan with separate dates for each action as the remedial actions may be completed before the corrective actions.

- Ensure the action fields do not limit your terminology, such as stating Corrective and Preventive Action. You cannot take Preventive Action on a problem that already exists. This error arises because designers did not consult ISO 8402 before releasing the product.

- Ensure you can link the findings to the schedule so that you can indicate which audits have been planned, conducted and completed.

- Ensure you can track progress without viewing every finding. Those tools that can give you a progress report are more suitable.

- Ensure the reporting features do not limit you to sending the report to the manager every time. There may be occasions when this is not appropriate.

- Ensure that the product allows for managers to review the findings and feed back comments should they disagree.

- Ensure you can enter data without releasing it so that you can return to it later to correct mistakes.

- Those products with links to document-change tools save time but ensure that a reference to the change is inserted so as to permit an audit trail afterwards.

## Performance assessment

### Performance assessment strategy

System audits gather objective evidence about the way the system is being implemented. They do not normally gather data to judge whether specified objectives are being achieved, although they should tell us whether the practices will enable objectives to be achieved. This is rather different. Performance assessment is concerned with measuring achieved performance against measurable objectives.

In assessing performance of the organization, data generated by the quality system needs to be gathered and analysed, and conclusions need to be drawn as to whether objectives have been met. In any process that is under control there will be checking points and action taken on the results. Unless the data generated by these processes is consolidated to permit measurement of the process, there is no clear measure of how well the organization is performing. For instance, all products may pass through the inspection processes and none may be delivered that do not conform to customer requirements. However, the reject rate in-process may be high, staff may have to work overtime to make up the losses and there may be a large consumption of material. The process can be inefficient and still yield conforming product. Performance assessment is needed to identify these inefficiencies and provide opportunities for improvement.

### Performance assessment process
The process is illustrated in Figure 8.7.

### Performance indicators
Without indicators, staff will measure all sorts of things: time to do X numbers of Y and ratios of A to B. Managers may command an analysis of a wide range of subjects. However, all the analysis should be focused. No analysis should be carried out unless there is a target to meet. Analysis for the sake of it is counterproductive.

The quality objectives provide the basis for performance assessment. However, the objectives may be defined in terms of overall performance and a bridge needs to be built between these general measures and the operational processes that produce the results. For instance, there may be an objective to reduce the level of customer complaints by 2% per quarter but you have some 2000 customers scattered across the globe with some 30 sales outlets. How are you going to measure performance? What constitutes a com-

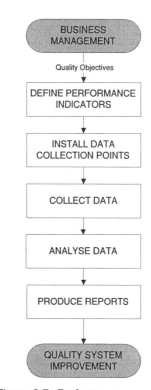

Figure 8.7 *Performance assessment process*

plaint in each of these sectors? It may not be the same. You may have an objective of continuous improvement but without defined indicators you will not be able to show how much improvement there has been and whether it is continuous. You may have many different processes each producing different products at different stages of development. A defect level for new products may well be higher than for mature products. The defect level for software will be measured differently than for hardware. Therefore performance indicators should be established for each product, process, service etc. to enable performance to be measured against measurable parameters. In this way, all your analysis serves a real need.

When you begin to determine the indicators you need, you may find important indicators arising for subjects that are not addressed by the quality objectives. If this is the case, then the quality objectives should be changed to reflect all the parameters that will give an indication of quality performance. Incidentally, you can have objectives for maintaining performance at a given level. You do not have to show improvement on every parameter. If a parameter has reached the optimum level, then continued monitoring should indicate whether the level is being maintained with any deterioration being a target for action.

*Data collection points*
Data collection points were mentioned in Chapter 5 on *System design* when analysing processes. You therefore need to review the flowcharts and decide where in the process the data will be generated to feed the Performance Assessment process. The data may well be raw data in a form that requires reduction and analysis to compare with the indictors. You may find that you have to modify records and reports to provide the right kind of information. For example, to meet a response time of two seconds at a help desk, the number of rings of the telephone needs to be logged and this can only be done automatically. The instructions for recording the rings and producing reports will therefore be in the telephone exchange software. Goods receiving defects may be logged on nonconformity reports that can be collected manually. If the data is entered into a database the right data fields have to exist to relate the data to receiving, the supplier and, on multiple sites, the location. The number of defects is no use without the source population. Even then, a rejected system costing £1m may be recorded as only one reject where 100 defective screws may be recorded as 100 defects. Such information neither tells us where to concentrate our efforts nor does it help in making comparisons.

*Collection of the data*
The data can be collected manually or automatically. However, unless you give instructions in the procedures for routing the data to the collection points it won't happen. In some cases a 'gofer' is dispatched to collect data every week or every month and this practice works in small companies. In large companies, electronic data collection is essential but there remains the issue of whether you are looking at the current data or historical data. The databases holding the data need to be defined and controlled such that the data is protected and current and no redundant data stored. Managers have a

habit of keeping old data because it was once useful to them. If you provide reliable up-to-date information, these managers will eventually throw away the old stuff.

With large organizations, data collection may be better handled on a functional or site basis so that there is data reduction before being passed to the centre for consolidation.

The frequency with which you collect the data depends on its rate of change. Some parameters may change daily, others weekly, some monthly and a few yearly. With some processes the data comes off the line to be analysed on the spot. If you are collecting data on warranty returns then hopefully there will be no difference day to day or week to week but over a period of months you may perceive a rate of change.

### *Data analysis*

Once you have collected the data you need to reduce it to a form in which it can be handled. Tons of computer printout paper are of no use to managers. They need to know what it means. Dots on X_R charts are of no use either, if the managers cannot understand them. Data analysis is a separate subject in itself but suffice to say that if it is not done well, the managers will not rely on the data and may well ignore it. The analysis should result in information that the manager will readily accept and follow up.

### *Production of reports*

Histograms, pie charts and graphs produced in colour look very attractive. Hand-written notes and scruffy calculations, whilst containing the same data, will not have the same impact. The reports need to get attention and so the data analyst may not be the same person who produces the reports. Skill is required to present the data of the company's performance in terms that will command action. The scales on the graphs need to be appropriate and not magnify a minor variation so that it appears more serious that it is.

The simplest way is to reproduce the quality objective at the top of the page and then show the actual performance in an appropriate manner, keeping it simple. Some managers like detailed reports. Others can't concentrate long enough to read them. Do your homework beforehand and pitch the reports at a level you know will attract attention.

## External audit

### *External audit strategy*

The external audit by the Certification Body provides an important input to the system evaluation process. Unless your customers require certification, there is no other external requirement to pursue registration to ISO 9000. Many companies see real advantages in having their quality system assessed by an independent body as it helps their marketing effort[5].

---

[5] See *ISO 9000 Quality Systems Handbook*, Part 2 Chapter 2.

Whilst many may believe that the external audit will tell them that their system is effective, it often only tells them whether or not they conform to the requirements of the standard. When you analyse the standard there are over 300 requirements in the 20 elements. It is not possible for the auditors to verify every single requirement in every relevant area it applies. Some certification bodies do claim to check every requirement and others admit to checking a sample. The problem arises because there is no definition on what a requirement is. In QS 9000 each 'shall' statement is checked, but these contain multiple requirements, all of which may not be checked.

In the law courts once the judgement has been made, a person cannot be tried for the same crime twice. This is why trials can take months or even years for all the evidence to be assessed. In a third-party audit, the auditors have more than one chance to prove noncompliance. The initial assessment will find the major problems and subsequent audits will continue to find many minor problems. Thus over a three-year period the whole system and every one of the 300 or so requirements will have been checked. This is the theory but auditing is not an exact science. Auditors differ in their competency and interpretation of the standard. A new auditor could arrive on site during a surveillance visit and find problems with the fundamental design of the system that should have been found on the initial assessment!

### External audit process
The planning, conduct and reporting of the audit are the responsibility of the third party. Nevertheless you have some responsibilities in this process that need to be defined. The External Audit Process is illustrated in Figure 8.8. The process may differ between certification bodies, as some review the manual on site. If you believe this to be more practical then insist that they do it. The process commences when system implementation is sufficiently advanced to provide a Quality Manual that reflects what you do or intend to do. The chart does not show your responsibilities during the audit and these and other matters will now be addressed.

### Choosing the assessment date
In general, certification bodies prefer that the system is implemented for about three months prior to the initial assessment. However, it depends where you start from as to when the clock starts ticking. If you had no semblance of a system in place, you will need three months of objective evidence that the system is being used. If some practices had been in place prior to launching the project, the changes to bring them into the system may be minor, making a proving period unnecessary. If you conduct audits for three months and then hold your first system review, you don't have to wait another three months and hold several reviews before the assessment. The initial assessment is a sample; therefore, even if the system is not fully in place, you can still proceed. There is a chance that the auditors will not select samples from the unproven practices. It is a little like a driving test. The learner driver may not have experienced every type of driving situation before taking the driving test but, nonetheless, is able to pass the test by demonstrating a capability to drive and understand the highway code.

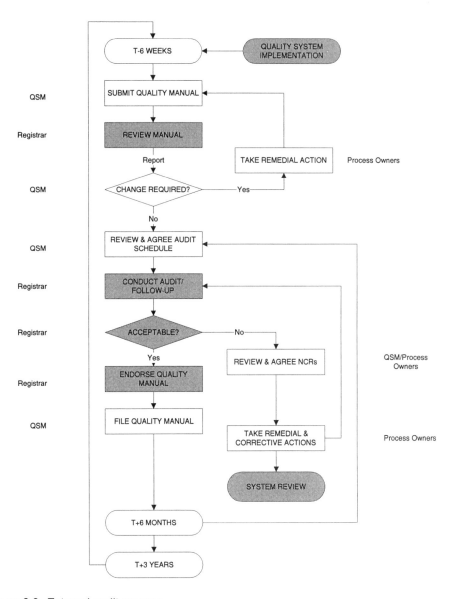

Figure 8.8 *External audit process*

The point to remember is that the initial assessment is a milestone on a journey towards world-class quality. When the certificate has been obtained, you have not finished developing your quality system – in fact you have only just started. The months and years ahead will cause you to refine the system continually. In some ways, system design may be complete but system development continues for ever!

## Submitting the Quality Manual

Depending on the policy of the Certification Body, at about six weeks prior to the scheduled initial assessment and subsequent re-assessments at three-yearly intervals, a copy of the Quality Manual should be submitted for the Certification Body to perform the documentation audit. The Policy Manual referred to in Chapter 6 substitutes for a Quality Manual but you may wish to provide other data, such as flowcharts, brochures and the ISO 9000 Exposition, if you have prepared one. If you have an on-line system then print out a copy that has date and revision status and a notice that it will be maintained. This may conflict with the banner you print on all documents indicating that printed versions are uncontrolled, but you can override this statement on the front sheet. You also need to exhibit its approval, which with electronic approval may not be obvious. Again a sheet containing the signature of the approval authority will suffice. Don't forget to retain a copy of the Quality Manual sent to the Certification Body as a reference document – it represents a baseline!

## Reviewing the Quality Manual

On receipt of the Quality Manual, the Certification Body will carry out a documentation audit and may request clarification or additional information. A record of any queries should be maintained so that it is readily available in case of dispute during the site audit.

## Taking remedial action

A record of any remedial action requested by the Certification Body should be maintained to bring the manual in line with the requirements of the standard. Changes should be processed using the approved change procedure and the affected pages resubmitted to the Certification Body for agreement if requested.

## Agreeing the schedule

The Certification Body will supply a draft assessment schedule showing the areas to be audited. An assessment schedule should be supplied for full assessments and surveillance audits. It is important to study this carefully as the person who prepared it may have misunderstood your organization and your system. A schedule showing audit by element should be questioned as to its effectiveness. The schedule should either address the functions or the processes. You also need to verify which locations will be subject to audit as adding them later may result in a reassessment.

You should consult the managers of the areas identified to determine their availability. In the event that an area targeted for audit is not practical for any reason (no staff on site, staff not available, conflict with other important visits etc.) the Certification Body should be advised accordingly, offering alternative dates, times or areas. Some managers may feel they are being neglected and others subject to undue attention. Ensure there is a good balance, since it may not serve your purpose to omit functions where you have had considerable difficulty. Remember that you have commissioned the audit and are paying for the service. He who pays the piper calls the tune!

Requested meeting rooms should be reserved and any arrangements for lunch initiated. When the schedule has been agreed copies should be distributed to all managers, including those not participating in the audit, to make them aware of a situation that may affect their effectiveness.

## Conducting the audit

The audit will commence with an Opening Meeting followed by the audit. At the end of each day or at the start of the next day a client feedback meeting may be held. The audit will be completed with a Closing Meeting.

### – The Opening Meeting –

Unless otherwise advised by the Certification Body, the following personnel should attend the Opening Meeting:

- The CEO or equivalent
- Process owners or functional managers
- Quality system manager (QSM) or management representative
- Nominated guides

The duty of staff present is to confirm that they understand the scope of the audit and the success criteria. They should also notify the Lead Auditor should any area designated for audit be unsuitable or key staff be unavailable.

### – Selection of guides –

Each auditor should be provided with a guide who should:

- Have knowledge of the areas/processes that are to be audited.
- Know who the key associated personnel are and where they are located.
- Have no other commitments during the audit.
- Have no responsibility in the areas being audited.

### – Duties of guides –

During the audit the guides should:

- Take the auditor to the place where he/she wants to go.
- Seek out the person with the responsibilities the auditor wants to interview.
- Introduce the auditor to the auditee.
- Co-operate with the auditor in seeking and agreeing objective evidence.

- Resolve problems in communication.
- Not answer questions put to the auditee.
- Not volunteer information unless requested.
- Tactfully advise the auditor should areas/questions etc. be outside the scope of the audit.
- Remain present at all times during an interview.
- Take notes of areas covered and note important findings.

**– Feedback meetings –**
Either at the end of each day or at the beginning of the following day, the Certification Body may wish to feed back the day's findings to the company. The Quality System Manager and guides should be present at this meeting. The process owners may attend at their discretion. Any findings requiring remedial action should be noted and their validity established. The Lead Auditor may choose to give a verbal report at the end of the day with formal written reports at the start of the following day in cases where audit extends to several days. Any reported problems should be noted and resolved with the managers concerned.

**– Agreeing audit findings –**
The QSM should liaise with the Lead Auditor and agree the audit findings on a daily basis ensuring that all have been agreed prior to the Closing Meeting. Before agreeing to the findings the QSM should establish that:

- The reported findings contain sufficient detail for remedial and corrective action to be taken.
- There is objective evidence that a requirement of the standard and the company's policies and procedures has not been met[6].
- Any classification of nonconformities is consistent with the definitions notified by the Certification Body.

Some auditors when finding a single lapse in following the system, record it and move on, in the belief that their role is to find nonconformity – this approach is not helpful to you. For the auditors to reach any sound conclusion they need to take representative samples. For example, a single instance of someone not knowing the quality policy does not indicate that the quality policy has not been understood at all levels in the organization. To test compliance with this requirement, the auditors have to take samples at each

---

[6] For further details on challenging the auditors, see *ISO 9000 Quality Systems Handbook*, Part 1 Chapter 5; for further details on dealing with challenging, see *ISO 9000 Quality System Assessment Handbook*, Chapters 4 and 5.

level in the organization. If they find it is not known at each level then a minor nonconformity may be justified.

A person needs to be knowledgeable about ISO 9000 to agree audit findings which is why this person is usually the QSM.

**– The Closing Meeting –**
Unless otherwise advised by the Lead Auditor, the same personnel who attended the opening meeting should attend the Closing Meeting. The duty of staff present is to confirm they understand the findings of the audit, agree to taking action and agree a date when proposals for remedial and corrective action will be submitted.

**– Appeal–**
In the event that agreement cannot be reached on the findings, the QSM should initiate an appeal in accordance with the Certification Body's procedures.

**– Recording of audit findings –**
The findings of the audit should be entered into the quality system to enable the actions to be tracked.

## *Submission of corrective action proposals*
Either at the Closing Meeting or prior to the date agreed, the process owners concerned should determine the appropriate action needed to respond to the findings. These should address as appropriate:

- The action needed to correct the problem
- The action needed to identify any other instances
- The action needed to alert staff to changes in advance of changes to policy/procedures
- The action needed to prevent a recurrence of the problem

Following agreement to the proposals, action to remove the nonconformity can commence. However, often only remedial action can be completed during the audit as corrective action requires more time to prove it has been effective. Corrective action can sometimes be initiated, such as changing documentation, but until the changes have been implemented and tested for compliance the action remains incomplete.

## *Taking remedial and corrective action*
The process owners should take the action proposed by the date specified and notify the QSM when the actions have been completed.

### The follow-up audit
On being notified that actions have been completed the QSM should initiate a follow-up audit to verify conformity. Unless otherwise notified, the QSM should advise the Certification Body of completion of the agreed actions.

In the event that the Certification Body requires a follow-up audit, separate to the routine surveillance visit, appropriate arrangements should be made.

### Endorsing the Quality Manual
On satisfactory completion of all outstanding actions, the Certification Body may request a copy of the Quality Manual for endorsement. A version of the Quality Manual should be provided that incorporates all the changes resulting from the audit. Following endorsement by the Certification Body of the paper copy of the Quality Manual, it should be filed and not changed, as it represents the baseline for the next audit.

### Changes to the manual
When a change is authorized to any document contained in the Quality Manual, it should be reviewed to determine the extent of deviation from the policies and practices endorsed by the Certification Body. Any change in policy or practice that alters the manner in which the company responds to a requirement of the standard should be notified to the Certification Body in the form of changed pages. It should be possible to process all other changes without such notification.


## System review

### System review strategies
The system review is the 'management review' referred to in clause 4.1.3 of ISO 9001. The term 'system review' is more accurate, as the review required by ISO 9001 is a review of the system not a review of management. The term 'management review' arises due to the review being required to be carried out by the executive management.

The purpose of the review is to ensure the continuing suitability and effectiveness of the quality system in satisfying the requirements of the standard, the stated quality policy and the stated quality objectives. The review should not be 'a meeting' because in reality you cannot review an entire quality system at a meeting. A way of combining the concept of a review followed by a review meeting would be to form a System Review Board (SRB) similar to the Material Review Boards (MRB) that are common in the defence industry.

### The System Review Process
An effective system review cannot be held until the system audit and performance assessment practices have generated sufficient data to draw conclusions about the effectiveness of the system. However, so that you are not found to be noncompliant you should conduct at least one review before the external audit. This first review may well be superficial

but as the system reaches maturity, the reviews will have greater impact. The system review process is illustrated in Figure 8.9. Data from both external and internal system audits and performance assessment feed the preparation process, together with statistics from the Quality System Improvement and Maintenance Processes and the Document Control Process. A System Review Board composed of the executive managers reviews the input data and determines if any changes are needed to the system to improve its effectiveness. The output from the system review produces records that are fed in, the Records Control Process from which actions generate maintenance, improvement or document change.

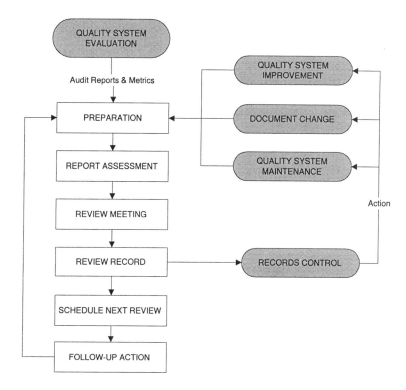

Figure 8.9 *Quality System Review Process*

### Preparation for the review
Data from several sources should be collected and should include:

- The system report from the Quality System Improvement Process, containing summaries of audit data, resource data, problem reports

- Performance data from the Performance Assessment Process, which should be summarized, indicating performance against the quality objectives

## 318 The quality system management process

- Quality system documentation data covering changes since the last review and summarizing the reasons for change,

- The results of a review of the Policy Manual, and any changes required summarized for discussion at the review meeting,

- Corrective action plans arising from the Quality System Maintenance Process, showing progress,

- Improvement plans arising from the Quality System Improvement Process, showing progress,

- Resource issues not yet addressed by improvement plans,

- Organizational issues not yet addressed by improvement plans.

The data should be compiled into a report that contains a summary page which addresses key issues to be resolved.

In large organizations, the preparation may be more easily handled if performed on a site or functional basis so that the QSM merely consolidates the data provided by others.

The report should be issued to the Quality System Review Board prior to a review meeting and in sufficient time for it to study it and review the practices under their control. In addition, an agenda for the review should be issued specifying the date, time, location, attendees and subject matter. A typical agenda is as follows:

1. Introduction and record of attendance
2. Record of last meeting
3. Actions from last meeting
4. Performance report
5. Audit report
6. Quality system status and pending changes
7. Policy review report
8. Resource issues
9. Organizational issues
10. Conclusions
11. Date of next meeting

## Assessing the review report
There is little point in the system review only being conducted by the Quality System Manager. The executive managers use the system and it is they who should review it for its effectiveness – but they are busy people and may expect to be provided with summaries to save time.

A System Review Board should be formed, composed of the executive managers and other managers who contribute to the management of the system. The composition will depend on which functions are within the scope of the quality system. However, all functions should be represented as they will invariably be affected by the quality policies or be required to service their implementation. On receipt of the system review report, the System Review Board should examine the data and establish that it is a true reflection of the status of the quality system. Any processes or procedures under their control should be reviewed to establish that they are suitable and effective and any issues to be resolved noted. This review can be delegated provided that the results are brought to a meeting of the System Review Board. The status of any outstanding problems should be established and data gathered to report the reasons for any delay in action.

## Review meeting
A review meeting is necessary to discuss the results of the system review – not to carry out the review. As stated previously, you cannot review a quality system at a meeting. The most you can do is examine the results of the review, reach conclusions and decide on the action to be taken.

All member of the System Review Board should be present at the System Review Meeting. The meeting should be led by the CEO or nominated representative. If the CEO does not wish to lead the meeting or even attend it, then it brings into question his/her perception of what the quality system serves to accomplish. The achievement of quality has to be led from the top of the organization. The quality system is the means provided by the organization to meet the quality policy and stated objectives. Therefore the absence of the CEO sends out the signal that he/she has more important things to deal with. Such things need to be put on the table so that you can judge whether the CEO has got the priorities right. One reason why any of the managers should be present is that the system review should commit managers to action. Policies will be established at the review. If these policies are not shared by all then their absence may prove counter-productive, leading to possible conflict later. They should attend the review meeting having received the review report and having had time to study it. However, it is not unusual for executives to arrive at the review meeting unprepared. If this happens then take appropriate corrective action and get it recorded in the records of the review!

In ensuring the effective implementation of the quality system the managers should ask:

- Does the quality system fulfil its purpose?

- Do the results of the audits indicate that the system is effective?

- Are procedures being used properly?

- Are policies being adhered to?

- Are the customers satisfied with the products and services we provide?

- Are the quality objectives being met?

- Do all the processes function cost effectively?

If the answer is 'Yes', your system is operating effectively. If your answer is 'No' to any of these questions, your quality system is not effective.

### The review record
The records of the system review include the review report and the result of the review meeting. A suitable format would be to record agreements and actions against each item on the agenda and append the review report as evidence. Minutes of the meeting alone are not evidence that the system has been reviewed. It is important to record conclusions from the review in terms of whether the system continues to be suitable and effective. If there are organization, technology or resource changes in the offing, then you need to decide whether the system will remain suitable following the changes.

### Review frequency
As a guide, reviews should be conducted more frequently when the system is immature – less frequently when it is stable with few changes. You may have a policy of reviewing the system at least once each year and this is satisfactory, but in reality you should be able to show that you are adjusting the review frequency depending on the maturing of the system and the results of the review.

### Follow-up action
The records of the review should be issued to the board members who should take any actions they have agreed to, using the established procedures. The QSM needs to check progress of these actions periodically as a safeguard against management inaction. Some managers will leave the meeting, put the papers in a file and do nothing until they are called to the next review; some managers act immediately by passing responsibility to others and then fail to check progress so that they are no better equipped than the managers who did nothing! Once the culture has been established, the independent checks will not need to be as frequent.

## System evaluation task list

1. Develop the audit management tools, procedures and forms.

2. Decide on your audit strategy.

3. Prepare the audit schedule.

4. Train the auditors.

5. Make management aware of what audits aim to achieve and how information will be gathered and used.

6. Assign auditors.

7. Plan, conduct and report individual audits.

8. Track remedial and corrective actions.

9. Analyse the results, determine root cause and plot trends.

10. Compile compliance tables showing the coverage of the audit programme and where the significant problems lie.

11. Form Corrective Action Teams to resolve problems.

12. Check that sensors are in place to monitor performance in key processes.

13. Determine and agree performance indicators with process owners.

14. Collect performance data to determine if quality objectives have been met.

15. Consolidate data to produce reports indicating the effectiveness of the system.

16. Provide reports to the system review.

17. Organize system reviews and agree the composition of the System Review Board.

18. Obtain management commitment to the system reviews.

19. Conduct system reviews and meet with management to agree results.

20 Record the results of the review and follow up on any corrective actions.

21 Arrange external audit.

22 Agree external audit schedule.

23 Take ownership of baseline Quality Manual for auditors to sign.

24 Co-ordinate on-site audit.

25 Deal with nonconformities as they arise.

26 Enter external audit results into database.

27 Organize follow-up audits.

## System evaluation questionnaire

1 Have you decided how you are going to determine the effectiveness of the quality system?

2 Are the tools in place for scheduling and reporting audits?

3 Have the auditors been trained in auditing and in the use of the tools?

4 Have you decided on the classification of audit findings?

5 Have you decided whether auditors will audit products and processes as well as systems?

6 Have you decided whether you will audit managerial decisions?

7 Has management been appraised of what to expect from the auditors?

8 Have managers been trained in the use of the corrective action tools?

9 Have you decided how many audits are needed to meet your audit objective?

10  Have you a sufficient number of trained auditors for the task?

11  After each audit do you know what was audited, what was checked and what the results were?

12  Are all audit results logged in the database?

13  Have the managers been given time limits to propose remedial and corrective actions?

14  Are the managers meeting their commitments?

15  Have you decided how you will measure the effectiveness of auditing?

16  Have you decided how you will present the results of audits to management?

17  Are you adjusting the audit frequency based on previous audit results?

18  Have you surveyed management reaction to audits and adjusted the strategy accordingly?

19  Are you retraining your auditors in response to negative feedback from managers?

20  Is there an escalation process in place to deal with problems which remain outstanding?

21  Have the performance indicators being used been agreed with the managers before reporting performance at the system review?

22  Have you decided who is to participate in the system review?

23  If the CEO cannot chair the meeting, have you established who will deputize?

24  Have you recorded the results of the system review of the quality system?

25  Have you recorded the agreements reached on current performance?

26  Have you decided whether the system will remain suitable after any planned changes are implemented in the organization?

27  Have you decided whether the system is effective and whether changes are required?

## System maintenance

### System maintenance strategy

Maintenance is concerned with retaining something in or restoring something to a state in which it can perform its required function. Quality systems comprise the organization, resources and processes as well as the documentation needed for achieving quality so you need to maintain more than the documentation.

In maintaining a quality system you need to:

- Keep customers satisfied with the products and service provided.

- Keep the quality system documents updated with the needs of the business.

- Keep copies of the documents updated with the latest amendments.

- Keep the policies and procedures up-to-date with the latest industry practices and technologies.

- Keep staff training up-to-date with current policies and procedures.

- Change policies and procedures to prevent the recurrence of problems.

- Keep the description of the organization, including the associated responsibilities and authority, compatible with the actual staff relationships and their responsibilities and authority,

- Keep the resources required to implement the policies and procedures compatible with the actual resources available.

In maintaining the quality system you are therefore doing more than maintaining pieces of paper. Often you are improving the system but by better control rather than by raising standards.

### System maintenance process

The system maintenance process is a process that enables the aspects addressed above to be carried out. The process is illustrated in Figure 8.10. Customer feedback (both negative and positive) is captured and routed either through the corrective action process or through records control to be reported to the System Review Board. Data from the System Implementation Process feeds Quality System Maintenance in the form of the

The quality system management process  325

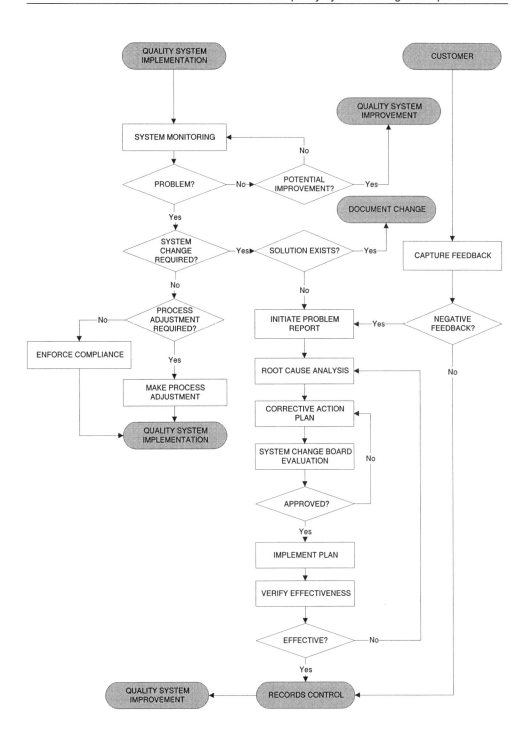

Figure 8.10  *System Maintenance Process*

actions and decisions taken by managers and staff. Some actions require no more than redirection or a process adjustment to restore the status quo. Others require change to the system to prevent recurrence. Some problems may be resolved through the Document Change Process and others may require more analysis through a corrective action process which may result in document changes, resource changes, organizational changes etc. Opportunities of improvement detected during system monitoring and the results of corrective action programmes feed the Quality System Improvement Process with data. This data can be analysed to find trends that may indicate a deterioration in standards and hence opportunities for improvement.

## System monitoring

The quality system should be owned by the managers. It is the system which enables them to achieve their objectives and so they should ensure it is used. Managers are either trying to maintain the status quo or change the status quo to achieve higher levels of performance[7]. It follows therefore that to maintain the status quo, managers must know what their organizations are achieving and take action should performance drop below expectations. In this way they maintain control.

In order to maintain system effectiveness, all managers should continually analyse process data, and observe the behaviour of their staff and the utilization of their resources in order to:

- Identify performance variation that impedes system effectiveness.

- Identify resource deficiencies and take action to prevent any adverse effect on system effectiveness.

- Identify demands that stretch the capability of the system.

- Identify organizational issues that impede system effectiveness.

- Identify variations between documented practices and work practices.

To gather this data you will need to plot results against targets for key product and process characteristics. In clause 4.9d of ISO 9001 process monitoring is required. Unlike the system audit process, which verifies that policies and practices are being followed, the system monitoring element should be focused on products, services and processes. Depending on your business, you may build such monitoring into the production processes and, in such cases, the system monitoring element will exclude product and process monitoring, concentrating instead on the soft issues such as organizational and resource issues.

[7] See *Managerial Breakthrough*, J. M. Juran (1964), Chapter 1.

Having identified any of the above variations, the manager should establish whether the variation can be eliminated by enforcing compliance or adjusting the process, or whether a system change is necessary. Process adjustments are not changes to the process but adjustments to the operating conditions to achieve the required results. A particular process change may well provide a greater range of adjustment so that the process remains capable.

In addition, the system should be kept in line with changes in the business, the organization, resources, technology, external requirements and changes in staff, and these are addressed below.

### Business changes
In order to keep the system up-to-date with the needs of the business you will need to review the system when changes occur in the business. This review may be carried out at the same time as the system review. As these reviews may be scheduled on a periodic basis, you should not allow the system to become out-dated whilst you wait for a meeting. The system should always reflect what you do and should remain ahead of actual practice rather than lag behind it. By considering the effect of business changes on the quality system at your weekly or monthly management meetings, you will provide safeguards against the system becoming unstable.

### Organization changes
A common failing of many quality systems is that the organization structure, job titles and responsibilities are out of date shortly after the documentation is issued. Managers often believe that the organization charts in the Quality Manual are there simply as a publicity aid and not as a definitive statement. Managers also prefer to be free to change their organization when it suits them and not to be constrained by a bureaucratic system. Most managers will announce a change in their organization, then rely on the Quality Manager to change the charts in the Quality Manual. To avoid conflicts you need a method whereby managers change the charts, then announce the changes in their organization, and not vice versa. Again, if you employ a quick change procedure, managers will find no advantage in by-passing the system.

### Resource changes
Policies and procedures, including the processes they define, require human, material and financial resources to implement. When you introduce the policies and procedures for the first time you need to take into account the resources that will be needed. It is of no use to issue a new procedure which requires new equipment, new skills and many more people if no one has made provision for them. Likewise, when procedures change you need to consider the impact on resources and when resources are reduced you need to consider the impact on the procedures. Managers may inadvertently dispose of old equipment or acquire new equipment without giving consideration to the procedures or instructions that specify the equipment. Some procedures may be designed around a certain facility or around a particular department, section or even a particular person or skill,

although every attempt to make them immune to resource changes was taken. In times of a recession certain pruning may have to occur which may affect the implementation of the procedures. You therefore need to be vigilant and identify the effects of these changes on your procedures and take prompt action to maintain them in line with current circumstances. Rather than dispose of procedures that have become obsolete due to such changes, archive them because you may be able to resurrect them when circumstances improve. Obsolete documents are also useful, to recall certain wording or phraseology when creating new ones, particularly where translated material is used.

### Technology changes

To keep your policies and procedures up-to-date with the latest industry practices you should provide a means of identifying new developments. This can be done by scanning journals, attending seminars and conferences, and generally maintaining an awareness of developments in quality management and technologies relevant to your business. When any new technology is to be introduced, the effect on the quality system should be included in the discussions before changes are made.

### Changes in requirements

The standard on which you based your system design may change and cause noncompliance. Some of the changes may be minor and require little adjustment. Other changes may be major and warrant a separate improvement programme. You need to monitor the standards to which you refer so that you are alerted to changes. The External Document Control Process should capture these and ensure action is taken.

### Staff changes

When you set up your quality system, as part of its implementation you should train staff in the application and use of the various documents. The system may not change as frequently as the staff so as new staff enter the organization or change roles, they need to be trained to carry out their jobs as well as possible. This training needs to be a continuous process if the standards of quality are to be maintained with a fluid workforce. You will therefore need a means of identifying when staff changes occur, so as to schedule their training. These training plans are as much a part of quality system maintenance as staff induction and development. Therefore, provision needs to be made in your procedures to ensure this occurs.

## Corrective action

### The nature of corrective action systems

Corrective action is action taken to eliminate the cause of a problem in order to prevent its recurrence. You can only take corrective action when a problem exists. The action you take to eliminate the problem rather than its cause is remedial action. You may have many remedial action mechanisms in place. The customer complaints process, the nonconforming product process, the design review process etc. all take action to correct the error but may not seek to establish the root cause and prevent its recurrence. One reason for this is that such action on each problem may not be appropriate. It may not be cost effective to chase through every problem on an individual basis. It is often better to collect the data, group like symptoms together and act upon the common cause. This is where a separate corrective action system is useful.

### Corrective action tools

You can provide separate corrective action systems or utilize a common system. Common systems work well with electronic data collection through a database. The Problem Management Database addressed previously in Chapter 4 can be used as a common database for all system problems if fields are provided to sort data by source and type.

### Identification of problems

Figure 8.10 shows that the problems requiring no change to the system are handled by line management without the necessity to record them. You may take the view that such records would be useful to detect trends and so they would, but before you consider recording such changes, talk to the managers themselves. They probably do not record every instance where they have to educate an employee in correct working practices and requiring them to do so may meet with considerable resistance. They may not record every adjustment they make to the process unless it is a mass production process where frequent adjustments signal other problems. What is important, however, is to record all changes to the system. Changes to resources are changes to the system; therefore any formal retraining either on-the-job or using structured courses would need to be recorded through the established training procedures. Other than these routines, if a solution already exists to the problem you can initiate a documentation change through the Document Change Process. For all other changes, a different route is required.

### Problem reports

A formal means to capture problems requiring corrective action is needed. With a Problem Management Database there will be a form to contain details of the problem, its source, location, the name of the originator of the report and classification of type and effect.

Problems requiring corrective action can be Actual Problems or Suspect Problems. A particular problem may not be isolated to the current incident as others from the same population may be scattered elsewhere, either within the organization or on other sites, including customer premises. In these other cases you don't know whether there is nonconformity, but they are suspect until proven otherwise.

An example of a Problem Report is provided in Figure 8.11.

| | | **PROBLEM REPORT** | PR/ |
|---|---|---|---|
| **Originator:** | | **Location:** | **Date:** |

| **Type** | Actual ☐ | Potential ☐ | Suspect ☐ |
|---|---|---|---|
| **Priority** | Urgent ☐ | High ☐ | Low ☐ |
| **Effect** | Critical ☐ | Major ☐ | Minor ☐ |
| **Category** | Safety ☐ | Environment ☐ | Performance ☐ |

**Incident** (What, when, and effect)

**Immediate action taken** (To contain effect)

**Remedial action** (To restore operations)

Approved . . . . . . . . . . . . . . . . . . . . . . (*Process Owner*)   Date . . . . . . . . . . . . . . . . . . .

**Root cause**

**Corrective action** (To prevent recurrence)

Approved . . . . . . . . . . . . . . . . . . . .   . . . . . . . . . . . . . . . . .   Date . . . . . . . . . . . . . . . . . . .
                Process Owner        QSM

**Verification**
That actions have been effective . . . . . . . . . . . . . . . . . . . . . . .   Date . . . . . . . . . . . . . . . . . . .
                                Process Owner

Figure 8.11 *Problem Report*

## Root cause analysis

The root cause of the problem should be determined using suitable disciplined methods. Disciplined methods are those proven methods that employ fundamental principles to reveal information. There are two different approaches to problem solving. The first is used when data is available, as is the case when dealing with nonconformities. The second approach is for when the data needed is only partially available.

The seven quality tools in common use[8] are as follows:

1   Pareto diagrams, used to classify problems according to cause and phenomenon

2   Cause and effect diagrams, used to analyse the characteristics of a process or situation

3   Histograms, used to reveal the variation of characteristics or frequency distribution obtained from measurement

4   Control charts, used to detect abnormal trends around control limits

5   Scatter diagrams, used to illustrate the association between two pieces of corresponding data

6   Graphs, used to display data for comparative purposes

7   Check-sheets, used to tabulate results through routine checks of a situation

The further seven quality tools for use when not all data is available[9] are:

1   Relations diagram, used to clarify interrelations in a complex situation

2   Affinity diagram, used to pull ideas from a group of people and group them into natural relationships

3   Tree diagram, used to show the interrelations among goals and measures

4   Matrix diagram, used to clarify the relations between two different factors (e.g. QFD)

5   Matrix data-analysis diagram, used when the matrix chart does not provide sufficiently detailed information

6   Process decision program chart, used in operations research

7   Arrow diagram, used to show steps necessary to implement a plan (e.g. PERT)

---

[8]  See *Kaizen*, Masaaki Imai (McGraw Hill, 1986).
[9]  ibid.

There are other techniques, such as force field analysis and the simple 'Why? Why?' technique, which often reveals the root cause of a problem very quickly.

Depending on the type of problem, it may be appropriate for a Corrective Action Team[10] (CAT) to be assembled to determine the root cause and prepare a Corrective Action Plan. This technique is useful when the problem spans more than one department or function. Representatives from each function can get together to resolve the problem.

### Corrective action plans

Once the root cause has been determined solutions can be proposed and tested and a Corrective Action Plan (CAP) put together for implementation. The plan may be contained on the Problem Report or in a separate document, depending on its size. If a separate document, it should be referenced on the Problem Report. The plans should indicate:

- The actions required

- The responsibility for each of the actions

- The dates by which the actions are to be completed

- The results expected from the actions in measurable terms

- The tests to be performed in order to verify the action fulfils its purpose

- The identity of the procedures to be followed or the methods to be used

- The records to be generated and maintained

The plan should receive approval from the System Change Board (SCB) as it will change the system.

It should not be presumed that corrective action plans will only address the procedural, product and process issues. If by asking 'Why? Why?' the root cause of the problem is found to be behavioural or cultural, the CAP has to address the action needed to remove the problem. This may be a very sensitive issue, particularly if the source of problem is a manager or even the CEO. Such plans need to be treated tactfully, otherwise they will not be well received and will be taken no further. It is this part of the quality system where the mechanized processes interface with the behavioural processes and TQM begins. An attempt is often made to use the CAP as a vehicle for behavioural change. This will fail if the author simply follows procedure. The Quality System Maintenance Procedure

---

[10] Otherwise known as 'QC circles' in Japan. See: *What Is Total Quality Control?*, Kaoru Ishikawa (Prentice Hall, 1985); *Kaizen*, Masaaki Imai (McGraw Hill, 1986); *Total Quality Management*, John Oakland (Butterworth-Heinemann, 1993).

therefore needs to recognize where this boundary will be crossed and should make provision for alternative methods of planning corrective action under such circumstances.

## System Change Board approval

In most quality systems changes are perceived as documentation changes and processed through the document change procedure. When a quality system has been designed using the systems engineering approach, one can no longer treat every change in isolation. A change to a task may affect other tasks, a change to a process may affect other processes and so on, potentially to affect the system design. There will also be changes that have no effect outside the document they impact. There is therefore a need for a two-level process for all changes: one for system changes and one for document changes. The document change process is addressed in Chapter 7. We will now address the system change process.

A system change can be classified as a change that affects the design of the system and any of the business or work processes. If we regard the documents merely as the vehicle used to describe the system and its components, we may change a document either in isolation or as a result of an associated system change. Minor document changes need only be presented to the document owner. Major document changes may also need to be presented to the associated process owner. Where a change affects the system as a whole or more than one process, a team of people needs to be brought together to agree the plans for change and to co-ordinate its introduction. We can call this team the System Change Board (SCB), similar to the Change Control Board or Configuration Change Board (CCB) in common use in system engineering business. If the term 'board' does not fit the culture then maybe the term System Change Panel may be preferable.

The SCB comprises the Process Owners and the QSM. It is a lower level board than the System Review Board (SRB); therefore, it does not have to include executive management. The role of the SCB is to evaluate proposed system changes, agree the implementation plans (corrective action or improvement plans) and obtain the resources necessary for their implementation. The SRB does in the implementation phase what the Design Review Board (DRB) did during development phase. At the final design review the DRB accepted the system design and established a baseline to which all subsequent changes would be referred. Between the external assessments, you need to keep a record of all quality system changes so that the auditors may determine whether the system remains compliant with the standard. It is likely that the auditors would wish to examine all changes but changes that meet the criteria for submission to the SCB would certainly be subject to scrutiny by the auditors.

The criteria for evaluating changes are covered in Chapter 7 on *Change evaluation*.

## Implementing the plan

The corrective action plan should be implemented as defined and any difficulty should result in a revision of the plan. However, this is not always necessary providing a record of the actions and decisions is maintained.

## Verifying effectiveness

When the corrective actions have been completed a series of checks should be undertaken to verify that the actions have been effective. It may not be possible to verify effectiveness merely by checking that the actions have been carried out as planned. Further data may need to be gathered and the trends plotted to show that the status quo has been restored.

## Records

The records generated through the corrective action process should come under records control to demonstrate that:

- The initial problem was detected

- The root cause was established

- Corrective action was planned

- The action has been carried out

- The action has been effective

ISO 9001 only requires records for the results of the investigation into the root cause, but clearly you need more than this to manage the process effectively.

## System maintenance task list

1. Allow the system to work.
2. Demonstrate commitment.
3. Test understanding regularly.
4. Gather data from using the system.
5. Use it to make decisions.
6. Either work the system or get the system changed.
7. Eradicate bad habits.
8. Revise documentation as needed.
9. Maintain the audit programme and management reviews.
10. Submit to surveillance visits.
11. Take corrective actions to prevent recurrence of problems.
12. Set up a System Change Board to handle all system changes.
13. Maintain the records.
14. Maintain the policies and procedures.
15. Don't allow deviations.
16. Document and implement – don't implement then document.
17. Eliminate informal practices.
18. Record and resolve problems.
19. Change practices under controlled conditions.
20. Monitor business changes.
21. Monitor technology changes.
22. Monitor staff changes.
23. Monitor organization changes.
24. Monitor resource changes.

### System maintenance questionnaire

1. Have you decided how and who will monitor the performance of the system?
2. Have you chosen the monitoring tools and trained the personnel in their application?
3. Have you installed the sensors to detect variance in performance?
4. Have you arranged for the collection and analysis of data?
5. Have you installed a problem management tool?
6. Is root cause analysis being carried out on detected problems?
7. Is the root cause analysis effective in identifying the true source of the problems?
8. Are corrective action teams organized for cross-functional problems?
9. Are corrective plans prepared for preventing the recurrence of problems?
10. Are corrective plans endorsed by management?
11. Has the impact of change on other processes been evaluated and appropriate action taken?
12. Are the reasons for abandoning any corrective action plans justified?
13. Is there an escalation process in place to bring in specialist and management support if needed?
14. Have champions been appointed to lead the improvement effort?
15. Are corrective action plans being expedited?
16. Have provisions been made and executed for verifying the effectiveness of agreed corrective actions?
17. Is the documentation always changed ahead of the practices?
18. Are staff trained in the use of new practices?
19. Are you supplied with information about the business, organization, staff changes and technology changes so that you can take action to maintain the system?

# System improvement

## System improvement strategy

There are those who claim that ISO 9000 does not address quality improvement but they would be wrong. Clause 4.14 is all about improvement. Improvement by better control is covered by clause 4.14.2 (Corrective action) and improvement by raising standards is covered by clause 4.14.3 (Preventive action). How you view these requirements depends upon what you regard as a nonconformity since these clauses are aimed at eliminating potential causes of nonconformity. They could have been written better but the intent is plain, especially when linked to the statement in the introduction to ISO 9001 which states: *The requirements are aimed primarily at achieving customer satisfaction by preventing nonconformity at all stages from design through to servicing.* The most important requirement is in 4.14.3b which requires the preventive action procedures to *include the determination of the steps needed to deal with any problems requiring preventive action.* This is far ranging and can be interpreted to apply to any threats facing the business from internal and external forces. With this in mind there can be no aspect of the business that is overlooked.

The improvement strategy therefore needs to approach the subject from two different viewpoints: the analysis of current performance to detect improvement potential and the analysis of future trends in the market place to detect improvement potential.

## System Improvement Process

The System Improvement Process follows the System Evaluation Process and takes input from both System Implementation and System Maintenance to provide data from which opportunities for improvement can be derived. The System Improvement Process is illustrated in Figure 8.12. Input data is collected from several sources to analyse and detect trends that require action to prevent problems. The outputs from the System Review Process and the Business Management Process are reports requiring action. These are subject to assessment to determine whether the action can be accomplished using established systems or whether research is necessary to come up with a feasible plan. The output from the process is a series of improvement plans, the status of which is reported to the System Review Board.

**338** *The quality system management process*

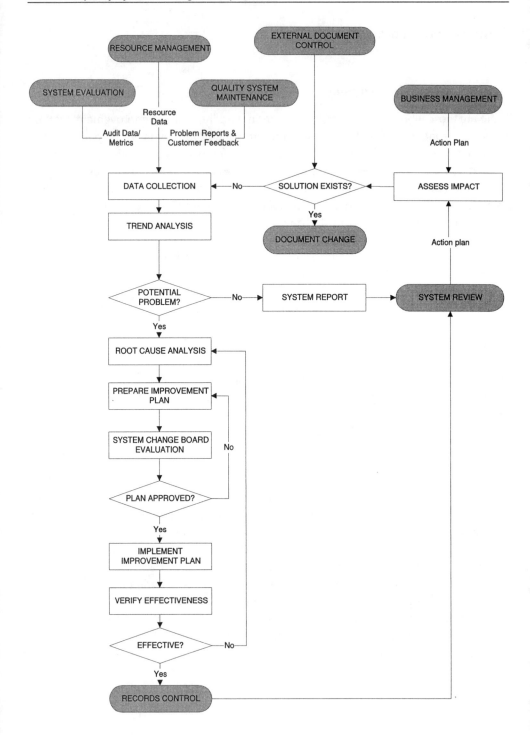

Figure 8.12 *Quality System Improvement Process*

## Data collection

Data from several processes needs to be collected:

- Audit data from the System Audit Process

- Audit data from the External Audit Process

- Problem reports from the Quality System Maintenance Process

- Customer feedback, including positive and negative reports

- Resource data from the Resource Management Process (specifically supplier problems, training statistics, equipment downtime etc.)

- Document change requests from the Quality System Maintenance Process or External Document Control Process

## *Trend analysis*

### – Audit data analysis –

The external audit data should be reduced to a form that will indicate the number of nonconformities and actions being taken.

The system audit data should be reduced to a form that will indicate both the coverage of the audit programme and the areas for improvement.

The data should be analysed to reveal trends in the concentration of nonconformities and their severity. An increase in documentation problems would suggest that staff are picking up bad habits and need training. A concentration of training issues would suggest that the Personnel Maintenance Process is not working effectively.

### – Problem reports –

Problem reports should be analysed to detect recurring problems which the corrective actions should have prevented. Also, where there is a concentration of problems in particular areas it is indicative of potential deterioration of standards.

### – Resource data –

- Supplier problems should be analysed to detect deterioration of standards.

- Equipment breakdowns should be analysed to detect wear-out of equipment.

- Maintenance data should be analysed to detect increases in corrective maintenance, being a sign that the preventive maintenance plan is not effective.

- Training programmes should be analysed to detect how many were completed as planned (indicating a possible lack of commitment or inadequate resources).

– **Change data** –

Approved document changes should be analysed to detect trends indicating:

- Deeper rooted problems with the documentation

- Processes exhibiting an unusual rate of change

- Functions exhibiting an unusual incidence of change requests

- Documents exhibiting an unusual number of changes, either very low or very high

– **Inadequate solutions** –

Where a change in external documents or an action arising from the performance review or system review cannot be implemented using existing practices, an improvement plan may well be needed. Before proceeding with any improvement plan, however, consideration needs to be given to other matters that may well influence the outcome. The changes in external documents or actions from internal review meetings may be reflected in current performance trends and either provide justification for pursuing or not pursuing a particular course of action.

## System report

The results of the trend analysis will indicate whether any action is needed. If no action is needed then the data can be presented to the System Review Board to show the current status of the system.

## Problem report

Should a potential problem be detected a Problem Report should be prepared. The same form as used for corrective action can be used so long as provision is made for indicating that a potential problem rather than an actual problem has been detected. The difference between an actual and a potential problem is one of timing. If the event has not yet occurred then it is a potential problem; if the event will occur in the time it will take to get the report out and attract management attention, then it should be treated as an actual problem.

## Root cause analysis

The same tools as used for determining the root cause of actual problems can be used. Instead of a Corrective Action Team to tackle the problem, you will need a Quality Improvement Team (QIT)[11]. Its brief will be longer term that the CAT. Some improvements take on the mantle of a major project, with a project manager and a multifunctional team.

## The improvement plan

The actual plan may not bear the title 'Improvement Plan' as it is too generic. The plan should carry a title appropriate to the project. Getting the management to think about all improvements as quality improvements is not easy. But if you have developed the system using the approach indicated in Chapter 5 and no provision has been made in the other processes for improvement, then quality system improvement is the process which management should use to develop the business. The data collection and analysis need not be performed only by one group. Any number of functions can apply the process and if all use the same database, then it will be simple for the QSM to determine what projects are being worked on. There can be no format for the improvement plan. If the plan can be laid out on the form then it simplifies matters. But if a separate document is needed then its content will be dictated by the type of problem. There are however, some aspects that will be common:

- A specification of the requirement or objective
- A strategy for achieving the objective
- A description of the organization required to implement the plan
- A statement of the work to be carried out
- The time-scales and costs
- The reporting structure
- The deliverables in terms of documentation and products

The plan may require few resources or significant resources. A project to go for ISO 9000 registration is such an improvement and requires considerable effort. The effort needed to improve product reliability may also be considerable and involve research into new technologies.

## System Change Board approval

The System Change Board established for dealing with system changes resulting from corrective actions should also deal with changes proposed for eliminating potential problems.

---

[11] For further details of QITs, see *Total Quality Management*, John Oakland (Butterworth-Heinemann, 1993).

## Implementing the improvement plan
The plan should be implemented as designed and, for large projects, updated as the project progresses.

## Verifying effectiveness
As with any change is it sensible to take measurements before and after. If the problem was detected using the methods previously defined then metrics will be available. As with the Corrective Action Plans, the verification of effectiveness cannot always be established immediately the actions have been completed. This is especially true about improvement plans. It may be months or years before the effects can be measured.

### System improvement task list

1. Set new objectives and targets
2. Develop annual improvement programmes
3. Look for potential problems – take preventive action
4. Seek suggestions for improvement
5. Simplify and optimize
6. Reduce complexity
7. Reduce the number of suppliers
8. Reduce variation
9. Reduce waste
10. Reduce time
11. Reduce error
12. Reduce job classifications
13. Reduce inspection
14. Reduce negative attitudes
15. Increase utilization
16. Increase training
17. Increase discipline & adherence to policy
18. Increase tidiness and cleanliness
19. Increase availability and retrievability
20. Increase motivation
21. Stabilize controls
22. Stabilize methods
23. Stabilize materials
24. Stabilize supplier performance
25. Stabilize processes
26. Stabilize the environment
27. Keep commitments
28. Keep records
29. Keep measuring performance
30. Keep analysing results
31. Keep auditing
32. Keep questioning
33. Keep reducing, increasing and stabilizing
34. Keep maintaining, improving and innovating

## System improvement questionnaire

1. Have you decided on your quality improvement strategy?

2. Do you have a policy for continual improvement?

3. Has every manager committed resources to the pursuit of continual improvement?

4. Are the tools in place to capture potential improvements to the quality system?

5. Have you provided training in the use of the improvement tools?

6. Is everyone aware of these tools and how to use them?

7. Are procedures in place to guide the improvement process?

8. Have you decided how improvement opportunities will be reported and to whom they will be reported?

9. Have you decided how you will organize improvement teams?

10. Do the improvement plans have management backing?

11. Are the reasons for abandoning any implementation plans justified?

12. Are the agreed improvement plans being implemented as agreed?

13. Are the plans effective in preventing the occurrence of problems?

14. Is progress on the implementation of the plans monitored and reported to the systems review?

15. Is there an escalation process in place to bring in specialist and top management support if needed?

16. Are you researching literature and using contacts to discover ways of enhancing your quality improvement programmes?

## System management organization

### The need

The organization that is put in place to manage the quality system development project will be unique. It has an objective to accomplish and when this is achieved its job is done and it can be disbanded. Why? Because its structure was formed to make a breakthrough and may no longer be appropriate for maintaining the quality system. During development you needed people with project management and system development skills – maybe people who could design databases, set up computer networks etc. Prior to certification, you need to put in place an organization that will maintain the system and this requires different skills. Although you may already have a Quality Assurance Function you may have assigned staff from this function to make up the team along with members of other functions. The existing function may not be equipped to taken on the role of quality system management. Conversely, the QA function may be well equipped to take on the role of system management – it may need some reorganization to do so. So what form should this organization take?

ISO 9001 requires a management representative to be appointed. Such a position was unnecessary during development. In addition, auditors are needed to audit the system; those that were trained during development are candidates for forming the audit team following hand-over to the line managers. The role of the Process Owner will continue and there is still a need for authors, but the Steering Group and Project Manager are no longer required. The Steering Group is replaced by the System Review Board made up of the executive managers (also known as the management review). The role of Project Manager is replaced by the role of Quality System Manager. Unless there is someone who will take responsibility for the maintenance and improvement of the system, it is likely that it will deteriorate in performance and fall into disuse. The system is dynamic and requires management to deliver optimum performance and hence its management has to be resourced. The consultant may of course be retained and can be used to provide training and further development advice. There are many opportunities beyond ISO 9000 registration, such as benchmarking, team building, problem solving and others where the use of independent advice may well provide added benefit to the organization. In small organizations, the consultant may be retained to carry out internal audits.

### The management representative

The term *management representative* is an invention of ISO 9000. In fact, other than in the title to clause 4.1.3, there is no other mention of a position with this name. It was also not the intention of ISO 9000 to force position titles upon companies but many companies have in fact created a new position or extended the role of an executive manager to carry out this role. The management representative is a role and not a position. Hence the person performing this role can have whatever title the company chooses.

Emphasis has been put in the standard on management appointing a member of its own management, indicating that the person appointed should have management status. This implies that the role cannot be filled by a contractor or external consultant. However, the appointment is a role not a full-time job. The person represents management to the Certification Body, to the customer and to suppliers on matters of quality.

Clause 4.1.3 states that the management representative *shall have defined authority for:*

a) *Ensuring that a quality system is established, implemented and maintained in accordance with this International Standard .*

b) *Reporting on the performance of the quality system to the supplier's management for review and as a basis for improvement of the quality system.*

There is a note which states:

> *The responsibility of a management representative may also include liaison with external parties on matters relating to the supplier's quality system.*

This requirement appears to be a contradiction in terms. Logically a representative carries the wishes of the people they represent to a place where decisions are taken that affect them – Members of Parliament, Union Representatives, Committee Members etc. The 'note' would appear to address the need for representation outside the business. Inside the business, the person represents management to the workforce but not in the same sense. The person carries the wishes of management (i.e. the policies) to the workforce so that the workforce makes decisions that take into account the wishes of management. However, the requirement matches more closely the role of a director rather than an external representative, as this person is not only representing management but directing resources in a way that will enable the company to achieve its objectives. What the organization needs is not so much a representative but a director who can represent management when necessary and influence other managers to implement and maintain the system. Such a person is unlikely to be under the direct authority of anyone other than the CEO.

To give this appointment due recognition, an appointment at executive level would be appropriate. The title chosen should reflect the position and as stated previously need not be a full-time job. Often companies appoint a member of the executive to take on the role in addition to other responsibilities. It could be Marketing, Sales, Engineering, Production or any other position. The notion that there has to be independence is one that is now dated and a reflection of an age when delivery was more important than quality. A person with responsibility for delivery of product or service also carries a responsibility for the quality of his/her actions and decisions. A person who therefore subordinates quality to delivery is unfit to hold the position and should be enlightened or replaced.

A solution is to appoint a Quality System Manager who may have other roles but if you already have a QA organization it would be logical to make the QA Manager the Quality System Manager or appoint a Quality Director who has a team in which one person is the Quality System Manager.

## The role of the Quality Director

The Quality Director is the top quality related position in the company. As previously stated, it does not have to be a full-time job but does involve several duties. In principle the Quality Director role is similar to the roles of the Financial Director, Security Director, Safety Director etc. It is a role that exists to set standards and monitor performance, thus giving an assurance to management, customers and regulators that specified objectives are being achieved. The role takes the title from the subject, which is vital to the survival of the company. If security is vital then a director is given the task of establishing security policy and putting in a system that will ensure security is not compromised. The security staff do not implement the system – that is the duty of all other managers. The same is true for finance, personnel, quality and any other critical success factor. If quality is vital to survival then it makes sense to appoint someone to direct the programme who will ensure quality is not compromised. As with finance, security and personnel these directors do not implement the policies, they regulate compliance. The other functional managers are appointed to deal with other factors critical to the company's survival and each is bound by the others' policies. This way of delegating authority works because it establishes a champion for each key factor who can devote resources to achieving specified objectives. Each manager is responsible for some aspect of security, finance, quality, personnel etc. Their responsibilities extend to implementing policy and achieving objectives. This means that the Production Director for example, is responsible for implementing the quality policy and achieving quality objectives within a system that is under the control of the Quality Director. Likewise the Production Director is responsible for implementing the design solution which is under the control of the Design Director.

If you were to make every manager responsible for setting policy, setting up systems and ensuring compliance then you would have as many quality systems, financial systems, security systems etc. as there were managers This is not an effective way to run a business. In such a structure, you would not have one company but as many companies as there were managers. If each manager is to serve common objectives, then we have to divide the objectives between them and permit one manager to impose requirements on other managers. This is what is known as *functional authority*.

The financial health of the company overall cannot be measured by each manager although individually they carry responsibility for the utilization of the financial resources within their own area. The financial health of the company can only be measured by someone who has the ability and authority to collect and analyse the data across all company operations. Man has a limited capacity for work. It is unreasonable to expect a person to be an expert in every subject. It is reasonable, however, to expect a person to accept the wisdom of experts in subjects in which he/she is not an expert. Hence the Quality Director should be the expert on quality just as the finance director is the expert on financial matters. This does not mean to say that they should take unilateral decisions – far from it – they should consult with the other directors to establish that the proposals will not conflict, thus enabling the organization to achieve any one objective without compromising the others.

This philosophy creates a role for a Quality Director whose responsibilities will include the following:

- Represent executive management with respect to the organization's quality system. (This means that the role interfaces with customers and suppliers on matters of quality.)

- Direct the development, implementation and maintenance of the quality system.

- Direct the evaluation of the quality system.

- Direct programmes that aim to improve the effectiveness of the quality system.

- Direct quality initiatives aimed at changing behaviour, image and values.

- Report on the quality performance of the organization.

### The product and service quality functions

The product or service quality functions such as Product Assurance, Software Quality Assurance, Reliability Assurance, Supplier Quality Assurance, Inspection and Test etc. can also be placed under the Quality Director or be dispersed to the line managers. For example, Product Assurance could be placed under the Product Development Manager along with Software QA and Reliability Assurance. Supplier QA could be placed under the Procurement Function and Inspection and Test under the Production Function. By placing these functions within the Quality Assurance function, you send out the signal that you don't trust the other functional managers to manage their processes. Dispersal of quality-related functions is therefore more in-line with current practice than their consolidation into central departments.

## Allocating work in the quality system function

The duties of the Quality Director can be delegated to two subordinate functions – that of a quality system development manager and a quality system evaluation manager.

Splitting development from evaluation provides a more easily understood division of work than division on the basis of task. It enables you to define objectives related to the processes rather than the individual and it provides for a cleaner decision-making process when deciding who to send to meetings, who to be involved with innovations or with resolving problems.

## The role of quality system development manager

The role of quality system development manager is to ensure that the quality system continually meets the needs of the business, its customers and its regulators. It is an innovative role requiring a person who is able to search for innovative solutions to problems and seize opportunities for beneficial change – a person who takes the broad view, sees the whole business and understands the way it operates. Some of the duties are:

- Keeping abreast of management system standards

- Networking with other companies, institutions and associations to develop mutually beneficial relationships

- Keeping abreast of the latest developments in management systems, practices, techniques and software

- Organizing quality management training and management awareness seminars

- Pursuing quality improvement initiatives and co-ordinating the development of new practices and processes through Quality Improvement Teams (QIT)

- Authorizing all quality system design changes

- Controlling the system documentation database, granting access, editing and deleting rights to staff

## The role of the quality system evaluation manager

The role of the quality system evaluation manager is to determine the effectiveness of the system in enabling the company to meet its objectives. The role is one of an investigator, requiring a person who is able to assess policies and practices, detect actual and potential problems, search for the root cause and initiate improvements: that is, a person who is analytical and methodical but who also takes the broad view and understands how the company operates. Some of the duties are:

- Keeping abreast of the latest development in auditing techniques, standards, practices and software tools

- Interfacing with external auditors and co-ordinating assessments on behalf of the organization

- Managing an internal quality audit programme to determine the extent to which policies and practices are followed

- Managing second-party audits on suppliers and determining the effectiveness of supplier quality systems

- Organizing auditor training and management awareness seminars

- Expediting corrective actions and organizing Corrective Action Teams (CAT) to discover root causes and initiate improvements

- Collecting and analysing performance data to determine the extent to which quality objectives are being achieved

- Controlling the problem management database and granting access, editing and deleting rights to staff

## Summary

This chapter has described the processes required to manage the quality system which collectively form a key business process. Often the individual processes are not consolidated in an organization. The system audit and review processes may be separated from the maintenance and improvement processes and these are often represented as several unconnected processes. Whilst there needs to be maintenance and improvement elements in other processes, by following through a logical sequence from system development, through implementation and onto maintenance and improvement, these common processes may be integrated to form a coherent subsystem. The links between the separate processes such as records control, document development and external document control were clearly indicated and the need for an external audit process demonstrated. A solution to the corrective action requirements of ISO 9001 was shown in the quality system maintenance process, where mechanisms for dealing with actual problems were described, taking data arising from internal monitoring as well as customer feedback. The preventive action requirements of the standard were shown to be represented by the quality system improvement process, thereby giving the correct focus to the requirements. Often preventive action is perceived to be the action taken to prevent the *recurrence* of an existing nonconformity, whereas it is actually meant to relate to the action taken to prevent the *occurrence* of a problem, as described in the *ISO 9000 Quality Systems Handbook*. It is not obvious from ISO 9001 that you need a system development process or a system maintenance process but the standard does require the system to be established and maintained and in this chapter I have shown how this may be accomplished through a series of processes.

Likewise, the standard does not require a system evaluation process but by combining the management review and internal audit requirements and taking into account the third-party audit process, the justification for creating a system evaluation process becomes clear. In many systems there will be processes that measure performance of particular products. In the process industries, the performance of particular processes will be measured. It therefore seems a small step to take to measure all processes and establish performance indicators to provide a sound basis for the measurements. Such measurements will provide an indication of how effective the system is in enabling the organization to meet its objectives. For this reason, performance assessment was judged to be part of the quality system evaluation process. All the audit and performance data needs to be reviewed; otherwise, any assessment of system effectiveness would be inconclusive.

Arguments were given for forming a System Review Board to review the system in place of the management review meetings that are a common feature of many quality systems. By treating the review in two parts (a review of a report followed by a meeting to discuss the results) the review should be more effective. In a similar manner, a System Change Board was suggested as a means of processing significant changes to the system, thereby formalizing what may have been rather informal processes and providing a focus for change. The Corrective Action Teams and Quality Improvement Teams that have arisen with TQM initiatives have been built into the model and shown to be part of an effective quality system.

Details were provided of the organization necessary to manage the system in order to complete the elements of quality system management. It should be appreciated, however, that the individual needs of organizations may dictate different solutions to those presented.

Whichever way you choose to manage the quality system, treating the processes that comprise quality system management as a key business process should improve awareness of how the system is being managed and bring added value to the organization.

## Chapter 9

# ISO 9001 Exposition

As ISO 9001 is not structured to meet any particular organization, a policy manual structured around the standard would not be user friendly, except perhaps to auditors; however, it is a convenient way in which to produce an Exposition. This chapter is devoted to such an Exposition but it is stressed that the *author* would not present a policy manual in this form. This is an Exposition, which declares how a company may respond to the requirements of ISO 9001. These are specimen statements written to suit a company that designs and produces electronic systems. The requirements of ISO 9001 are presented as questions in italics. The response provides solutions to the requirements.

Many quality manuals merely paraphrase the requirements of the standard. It is an approach that is recommended by the North American National ISO 9000 Users Support Group[1] and with which some Certification Bodies agree. It is an approach with which I disagree, primarily as paraphrasing the requirements is not the same as responding to the requirements. There is little use for a document that paraphrases requirements without indicating how an organization intends to meet the requirements.

The company can achieve the same result by stating in their quality policy that they will meet the requirements of ISO 9000. I also disagree that a manual presented as described in Randall's book will be of any use to the organization it represents. I don't see the difference between making a general statement that you will comply with the standard and rewriting it in a company document. If you are going to structure your quality manual so as to follow the requirement headings in the standard, then you should at least provide your solutions to the requirements: that is, your intentions or your response.

This chapter shows you how to respond to the requirements of ISO 9001 but you will observe that the statements are phrased as 'will' statements and not 'shall' statements. The manner in which this chapter is structured will enable you to create your own policy manual by rephrasing the statements as policies: that is, make them 'shall' statements. You can then leave the statements structured around the headings of the standard or

---

[1] Extract from *Randall's Practical Guide to ISO 9000*, Richard C. Randall (Addison-Wesley Publishing Company, 1995).

order them around your business processes – an approach I strongly advocate, as it will mean more to the employees. Alternatively, you can produce an Exposition just like this one and include it alongside your policy manual to indicate your response to the requirements. In such cases, the Exposition will not be a working document but a reference document, for use by auditors and others wishing to relate the standard to your business.

## Management responsibility

### Quality policy

(a) *In what documents do you define your policy for and commitment to quality?*

The company's philosophy and hence commitment to quality is declared in a Policy Statement signed by the Managing Director. The policies for enabling this philosophy to be maintained throughout the company's operations are documented in a Quality Manual, approved by the management, and copies are made available to all staff.

(b) *In what documents do you define your objectives for quality?*

In order to maintain established quality standards and facilitate continuous improvement in the quality of the company's products and services, quality objectives are established each year and a programme implemented to achieve them. The programme and resources to carry it through are authorized by the Managing Director and managed by the quality assurance authority.

(c) *How do you ensure that the quality policy is relevant to your organizational goals and the expectations and needs of your customers?*

The quality policy is developed in conjunction with other organizational goals, periodically reviewed by the management team against the market analysis of current customer expectations and needs, and amended as appropriate.

(d) *How do you ensure that your policy for quality is understood at all levels in the organization?*

These policies are developed in conjunction with the staff who will implement them and only published once they have been agreed and understood. Managers coach their staff to ensure understanding and provide help and advise in the implementation of the published policies and procedures.

(e) *How do you ensure that your policy for quality is implemented at all levels in the organization?*

Procedures and standards are developed to implement the agreed policies, referenced within the Policy Manual and subject to a comprehensive programme of independent audits.

(f) *How do you ensure that your policy for quality is maintained at all levels in the organization?*

The Policy Manual is the only vehicle for promulgating policy in the company. These policies are periodically reviewed by management and changes made to ensure that they remain current and relevant to the business.

## Organization

### *Responsibility and authority*

(a) *In what document do you define the responsibility and authority of all personnel who manage, perform and verify work affecting quality?*

The responsibility and authority of all personnel is defined within the procedures that apply to the operations they perform. In addition, the responsibilities, authority and accountabilities for those holding specific positions or carrying out a particular trade or profession are defined in Job Profiles.

(b) *In what documents do you define the interrelation of all personnel who manage, perform and verify work affecting quality?*

Each of the company's procedures indicates the interrelation of personnel in performing specific tasks or operating specific processes. In addition, Organization Charts are maintained to depict the reporting channels and the location of staff within the company where staff are located off the main site.

Where the policies and procedures indicate that actions and decisions are to be taken by certain authorities, the organization to which this authority has been delegated for a particular contract is specified in the Quality Plan.

(c) *How do you ensure that, when needed, personnel have the organizational freedom to identify quality problems, provide solutions, initiate action to prevent their recurrence and verify the implementation of solutions?*

All staff have the freedom to identify problems and initiate action to prevent their recurrence but only those with responsibility for certain results have the authority to determine how those results are to be achieved or improved. In the event of a member of staff identifying either an actual or a potential problem for which they are not responsible, the problem and any feasible solutions are reported to their manager for consideration. Any person neglecting to take advice given by a qualified person will be held accountable to their manager.

The personnel assigned to investigate particular problems, implement and verify solutions are identified in forms or reports that relate to the problem and these documents retained in the form of quality records.

(d) *How do you ensure that those responsible for results have the organizational freedom necessary to control processing, delivery or installation of product?*

Personnel responsible for releasing products or services to the customer or authorizing installation to commence are responsible for ensuring that all previously reported and documented problems have been resolved to the satisfaction of those concerned before commencing delivery or installation.

## Resources

(a) *How do you identify resource requirements?*

An estimate of the manpower, material, facilities and plant needed to execute a particular contract is established, documented and agreed with senior management prior to submission of any tender, bid or offer. The estimate includes the resources to manage and carry out the work required and in addition the resources required to verify that the work has been completed in accordance with the contractual requirements.

Annual resource budgets are prepared by each manager for his/her department that include all manpower, material, facilities and plant needed to meet the company's objectives including the resources needed to maintain the quality system and either carry out or participate in any audits, reviews and improvement programmes.

(b) *How do you ensure that adequate resources are provided?*

The agreed resources programmes are implemented through a sanction mechanism which prioritizes their acquisition in line with available finances. In the event of resources not being available to meet demands, temporary changes are made to the objectives and plans until the resources become available.

On receipt of either an invitation to tender or a contract, the documentation is reviewed to identify any verification requirements that may be imposed by the customer. Following determination of the work needed to satisfy the customer requirements and provide the required products and services, the aspects requiring verification will also be identified and the results documented in the form of plans and specifications that will govern the conduct of work under the contract.

(c) *How do you ensure that trained personnel are assigned for management, performance of work and verification activities including internal quality audits?*

Managers maintain records of staff skills and ensure that only adequately trained personnel are assigned to carry out any activities assigned to them.

## Management representative

(a) *Whom have you appointed to ensure that the requirements of ISO 9000 are implemented and maintained and how do you ensure that this person remains a member of your own management?*

A position of Quality System Manager has been established and maintained and filled by a permanent member of staff appointed by and responsible to the Managing Director.

The Quality System Manager has the delegated authority of the Managing Director to represent the company regarding the quality system employed to ensure that its products and services meet customer requirements. He or she is responsible for ensuring that the quality system meets the conditions required to maintain registration to any external quality system standard relevant to the business.

(b) *How do you ensure that the management representative has defined authority for ensuring that quality system requirements are established, implemented and maintained and who reports on quality system performance to the management review?*

The Quality System Manager has the delegated authority of the Managing Director to:

- Manage the design, development, implementation and evaluation of the quality system including the necessary resources.

- Determine whether proposed policies and practices meet the requirements of the standard, are suitable for meeting the business needs, are being properly implemented and cause noncompliances to be corrected.

- Determine the effectiveness of the quality system.

- Report on the quality performance of the organization.

- Identify and manage programmes for improvement in the quality system.

- Interface with the appointed Registrar.

## Management review

(a) *How does your management ensure the continuing suitability and effectiveness of the quality system in satisfying the requirements of ISO 9001 and the company's stated quality policy and objectives?*

Procedures are employed to collect and analyse objective evidence on the effectiveness of the quality system. This includes the results of internal and external audits, customer feedback, contract performance data, nonconformance data and problem reports. The results of the analysis are reviewed by the management team at periods determined by the Quality System Manager but no less than annually. The management team establishes whether the quality system is effective in enabling the company to achieve its quality objectives and maintain its quality philosophy and satisfy the relevant requirements of pertinent quality system standards. The team will also determine what action if any is needed to improve the effectiveness of the quality system.

(b) *Who is defined as management with executive responsibility?*

The managers with executive responsibility who report directly to the President are:

- Managing Director
- Finance and Administration Manager
- Sales and Marketing Manager
- Production Manager
- Chief Designer
- Quality System Manager

(c) *What records are maintained to demonstrate that your quality system is effective?*

The results of quality system reviews are recorded together with the assignment of responsibility for any corrective or preventive action required and retained in the form of quality records.

## Quality system

### Purpose and scope

(a) *What is the purpose of the quality system?*

The quality system has been developed for the purpose of:

- Ensuring that products and services provided to customers meet their requirements

- Ensuring consistency in the manner in which work is executed

- Ensuring that processes are repeatable and predictable

- Enabling the company to create and retain satisfied customers

- Improving efficiency, reducing operating costs and minimizing unproductive time

(b) *What is the scope of the quality system?*

The quality system applies to all revenue earning operations of the company except those concerned with site maintenance, company finances, catering, transport and staff remuneration and welfare. It applies to the design, development, procurement, production, installation, delivery and servicing of the company's product and services, to the components of those products and services and to any equipment used in conjunction with such products and services.

## Quality system development

(a) *What documents constitute the quality manual?*

The Quality Manual comprises the following documents:

- Policy Manual

- Quality system flow diagrams

- Index of control procedures

- ISO 9001 Exposition

(b) *In what document is the outline structure of the quality system defined?*

The quality system documentation structure is defined in the Policy Manual which is summarized as follows:

Documented control procedures have been prepared to control product and information as it passes through processes. Documented standards have been prepared to define the acceptance criteria for judging the quality of a product, service, process or document. Documented operating procedures have been prepared to convey requirements for performing a single task where the results are not dependent on skills.

(c) *What document covers the requirements of the standard?*

The Policy Manual covers the applicable requirements of the standard to the extent that they apply to the operations of the company. A Compliance Matrix indicates the relationship between the sections of the Policy Manual and the clauses of ISO 9001. This Exposition provides a response to each requirement of the standard.

(d) *What document includes or makes reference to the quality system procedures?*

The policies contained in the Policy Manual reference the applicable control procedures. An index of control procedures is included in the Quality Manual and the control procedures reference all other quality system procedures.

(e) *How do you determine the degree of documentation required?*

Documented procedures have been prepared to implement policies and quality system requirements where the results required rely on the combined efforts of various people carrying out a number of activities in a prescribed sequence.

(f) *How do you ensure that the quality system procedures that have been prepared are consistent with the requirements of ISO 9001 and your quality policy?*

The quality assurance authority will establish that all quality system procedures, standards and guidelines satisfy the requirements of ISO 9001 and the quality policy prior to their release for use.

(g) *How do you ensure that documented procedures are effectively implemented?*

The implementation of new quality system procedures is planned so that their introduction takes place effectively.

Prior to introducing a new policy or practice those who will be affected by it are notified as to what will be expected of them and appropriate training provided.

Where possible, old practices will not be removed until the new practices are operating effectively and the controls to verify effective operations have been established.

All new practices are monitored closely by the managers concerned to gain confidence in their effectiveness.

## Quality system maintenance

(a) *How is the quality system maintained?*

Consequent to changes in business needs, industry practices, technologies or external quality system standards, the quality system is reviewed and where necessary changes made to the documented policies and procedures. Whenever changes are proposed to the organization structure or resource levels, the policies and procedures are reviewed and, if affected, changed prior to or promptly after implementation of such changes.

In order to maintain quality standards, understanding, skills and knowledge when staff are redeployed, retraining is scheduled and implemented so as to minimize any disruption caused by the change.

## Quality planning

(a) *What manner do you define and document how the requirements for quality will be met?*

The quality system defines and controls the preparation, use and maintenance of those documents that describe how the required features and characteristics are to be designed and built into the company's products and how the achievement of these features and characteristics is to be verified.

(b) *How do you ensure that quality planning is consistent with other requirements of the quality system?*

Quality planning is undertaken in conjunction with other planning activities as directed by quality system policies and procedures.

(c) *How do you determine whether a quality plan is required for products, projects or contracts?*

A separate quality plan is prepared to tailor the quality system to a specific product, project or contract and define the specific responsibilities, activities and equipment required to implement specific product, project or contract requirements.

(d) *How do you identify and acquire any controls, processes, inspection equipment, fixtures, total production resources and skills that may be needed to achieve the required quality?*

On receipt of a contract the controls, processes, inspection equipment, fixtures, production resources (including equipment, plant, environment, machines, materials, processes, skills, documentation and utilities) required to execute the work is determined and specified in a contract-specific quality plan and the plan approved by the Production Manager and quality system prior to the commencement of the work concerned.

(e) *How do you ensure that the design, production process, installation, inspection and test procedures and applicable documentation are compatible with the specified requirements?*

In order to ensure compatibility between contractual requirements, company specifications, procedures and process capability, reviews are undertaken at each document release stage.

(f) *How do you identify whether any quality control, inspection and testing techniques and instrumentation require updating to meet specified requirements?*

Contracts, the associated documents and any specifications produced by the company to implement the requirements are assessed and any changes needed to the existing quality controls, inspection and test techniques and instrumentation determined. Where changes or additional provisions are necessary they are specified in the quality plan and arrangements made for their acquisition prior to implementation.

(g) *How do you identify measurement requirements involving a capability that exceeds the known state of the art in sufficient time for the capability to be developed?*

The contractual requirements and any company specifications produced to meet them are assessed for any measurement capability that exceed the state of the art. If such a capability is identified the customer is notified and the contract re-negotiated or the parameters relaxed.

(h) *How do you identify verification requirements?*

On receipt of either an invitation to tender or a contract, the documentation is reviewed to identify any verification requirements that may be imposed by the customer. Following determination of the work needed to satisfy the customer requirements and provide the required products and services, the aspects requiring verification are identified, the stage of verification determined and the results documented in the form of plans and specifications that will govern the conduct of work under the contract.

(i) *How do you clarify standards of acceptability for all features including those containing a subjective element?*

It is encumbrant upon the company to determine the customer requirements and ensure they are understood, achievable and verifiable using the resources at the company's disposal. Where the customer has not adequately specified the acceptance requirements and declines or is unwilling to be specific he is advised that the company standards will prevail.

(j) *How do you identify and prepare any new quality records to meet specified requirements?*

The contractual requirements and the plans prepared to execute a contract are assessed and the quality records required identified and listed in the quality plan. Any new records required are designed and implemented using quality system procedures.

## Contract review

### Review procedures

(a) *What procedures have been established and documented for contract review?*

Contracts and tenders are reviewed in accordance with the Contract Acquisition Procedure.

(b) *How do you co-ordinate contract reviews?*

For all contracts other than repeat orders, the contract authority liaises with the managers who will carry out the work required and obtain their agreement to the conditions. The Quality System Manager or his nominated representative will participate in these discussion as necessary. Agreement from these managers is obtained prior to recommending acceptance of the contract to the Managing Director.

### Review

(a) *How do you ensure that tenders are reviewed before submission?*

On receipt of an enquiry or invitation to tender, the customer requirements are examined by the contract authority to establish that they are adequately defined and that the company is able and willing to meet them.

(b) *How do you ensure that contracts are reviewed before acceptance?*

Contracts that were not subject to tender action, are reviewed by the contract authority and, if acceptable, authorized for processing. The company will not enter into contracts that will result in a breach of national and international legislation. Unless otherwise negotiated with the customer, the company's standard conditions of sale will apply.

(c) *How do you ensure that requirements are adequately defined and documented before accepting a contract?*

In order to establish that the contractual requirements are adequately defined and documented, a standard check list is employed which details all the conditions that need to be satisfied for the company to estimate the resources required to execute the work. Any additional details required are obtained from the customer where possible. In the event that such information cannot be obtained, the offer will define any assumptions that have had to be made.

(d) *How do you ensure that requirements differing from those in the tender are resolved before accepting a contract?*

On receipt of an order or contract for work for which the company has previously tendered, the contract and associated documents are examined to establish that there have been no changes that would invalidate the offer. In the event of there being such differences, the contract will not be accepted until re-negotiated.

(e) *How do you ensure that you have the capability to meet the contractual requirements before accepting a contract?*

A record of the company's current capabilities is maintained and this is consulted when determining whether the company will be able to execute the work specified. Where additional capability will be needed, the Managing Director determines how this capability will be acquired.

## Amendment to contract

(a) *How are amendments to contracts made and correctly transferred to the functions concerned?*

On receipt of a contract amendment, the contract authority informs those functions who participated in the contract review and obtain their agreement to the amendment. Any change which will affect the validity of the original quotation is costed, agreed with the customer and the appropriate plans changed to implement the new requirements.

## Records

(a) *In what documents do you record the results of tender reviews and contract reviews?*

On receipt of an enquiry, invitation to tender, order or contract, the salesperson records the action taken and the factors which led to the decision. Any correspondence, meetings or written comments associated with the potential order is traceable to the initial customer contact and retained in the form of quality records.

(b) *Where are the channels of communication and interface with the customer's organization defined?*

The quality plan or the contract specifies the names and responsibilities of the key personnel who will liaise between the company and the customer in the execution of the contract.

# Design control

## General

(a) *What procedures have been established and documented to control and verify product design?*

The following procedures are employed to control and verify product design:

| | |
|---|---|
| CP | Project Management |
| CP | Product Development Planning |
| CP | Systems Engineering |
| CP | Hardware Design |
| CP | Applications Software Design |
| CP | Product Development Tests |
| CP | Software Integration and Test |
| CP | Configuration Management |
| CP | Design Drawings |

(b) *How do you control and verify product design?*

Procedures are established, implemented and maintained which ensure that all product and service design activities are conducted under controlled conditions that result in designs that have been proven compliant with agreed design requirements.

Each new design is classified as a project. A project manager is appointed with responsibility and authority to direct the efforts of all those to whom design tasks have been assigned and to enforce compliance with the agreed design requirements.

## Design and development planning

(a) *In which documents have you identified the responsibility for each design and development activity?*

A design authority is appointed for each product and service which the company supplies. The design authority is the organization responsible for the detail design of the product or service and is authorized to certify that the design complies with the agreed requirements.

Design and development plans are produced for each product or service to be developed and are approved by the design authority before commencement of the design activities to which they relate. These plans define the design activities and the organizations responsible for carrying them out.

(b) *How do you ensure that design and verification activities are planned and assigned to qualified personnel equipped with adequate resources?*

The work required to create a design that meets the agreed requirements is divided into work packages. Estimates of the resources required are obtained from the relevant department. Where the company does not possess the necessary resources to carry out a particular work package, work may be subcontracted or labour bought in by implementing the relevant procurement policies.

(c) *How do you ensure that the design and development plans are updated as the design evolves?*

The design and development plans are reviewed at regular intervals. The plans are updated to reflect progress, changes in the design approach and the allocation of responsibilities.

## Organization and technical interfaces

(a) *How do you identify, document, transmit and regularly review the organizational and technical interfaces between different design groups?*

The design and development plan defines the relationship between all organizations that participate in the design. Documented work instructions are used to convey design requirements between internal organizations. Design activities to be carried out by subcontractors are specified in a statement of work that forms part of the subcontract.

Physical and functional interfaces between parts of the product or service are specified in interface specifications either as part of the design requirement specification or as a separate document. These are subsequently conveyed to all the organizations that need the information to produce the design.

All organizations participating in the design are provided with a current and complete list of the governing design and interface specifications.

The interfaces between design organizations and the technical interfaces between designs are regularly reviewed as the design progresses. Alterations are made as necessary to overcome weaknesses.

## Design input

(a) *How do you identify, document and review design input requirements including applicable statutory and regulatory requirements?*

A design requirement specification is either obtained or produced which defines all the features and characteristics that the product or service is required to exhibit to satisfy customer needs and expectations.

Any statutory and regulatory requirements pertaining to the operation and use of the company's products and services within the country to which it will be supplied are identified. These requirements are identified at the commencement of design and included in the design requirements specification.

(b) *How do you ensure that the selection of design input requirements is reviewed for adequacy?*

The specification is thoroughly reviewed by the design authority prior to commencing detail design.

(c) *How do you resolve incomplete, ambiguous or conflicting design input requirements?*

In the event of the design requirement specification containing ambiguous, conflicting or inadequately specified requirements, the nature of the inadequacies are documented and reported to the project manager for resolution with the customer prior to contract acceptance.

(d) *How do you ensure that design inputs take into consideration the results of contract reviews?*

Where a contract requires design work, the design authority participates in the contract review, receive a record of the results and ensure that the agreed requirements are promptly incorporated into the design requirement specification.

## Design output

(a) *How do you document design output requirements?*

The design solution is documented in a product or service specification in terms of all the features and characteristics that reflect a product or service which will satisfy the agreed requirements.

(b) *Where is the evidence that design output requirements can be verified?*

All calculations and analyses made to size components and to determine or predict characteristics are recorded in a design file together with any assumptions made in a manner that will facilitate their retrieval and verification should the need arise. Calculations and analyses that impinge upon human safety are certified by a second qualified person having no responsibility for the particular design.

(c) *How do you ensure that the design output contains or references acceptance criteria?*

The characteristics of the product or service is expressed in terms that are measurable using current technology. The criteria to be used to judge the acceptability of features and characteristics are defined in verification specifications either as part of the product specification or as separate documents. Metric units are used for all dimensions unless otherwise specified by the customer. External interfaces adhere to appropriate national and international recognized standards where available.

(d) *How do you identify those characteristics that are crucial to the safe and proper functioning of the product?*

Any characteristics that are crucial to the safe and proper functioning of the product or service are identified in the product specification as safety-critical. Means are provided to prevent or contain any hazards or alert operators should improper functioning be imminent.

(e) *How do you ensure that design output documents are reviewed before release?*

All design documentation that is to be used for building, procuring, operating, maintaining product etc. is reviewed and approved before being released to the design review board.

## Design review

(a) *How are formal design reviews planned, conducted and documented?*

Design verification is carried out by formal design reviews convened at discrete stages throughout the design process.

The stages at which the design reviews are to be held and the review requirements are documented either as part of the design and development plan or in the quality plan.

Each design review establishes that the design satisfies the requirements for the particular phase of its evolution and that the provisions for the subsequent phase are adequate. The designers present documentary evidence to the design review board demonstrating achievement of the prescribed requirements and fulfilment of the statement of work. This documentary evidence may be obtained as a result of analyses, calculations, simulations or by test, inspection or demonstrations on prototype models representative of the production version.

(b) *How are the results of design reviews recorded?*

The results of design reviews are recorded in terms of a definition of the criteria against which the design was reviewed, a list of design documents which were reviewed, the decision of the review board and a statement of any corrective actions required.

## Design verification

(a) *How do you ensure that design output meets design input requirements?*

No design is released for production or use until it has been demonstrated as meeting the agreed design requirements and certified acceptable by the design authority at a design review.

(b) *How are the design verification measures documented?*

The design verification provisions are documented either as part of the design and development plan or in a separate design verification plan.

Design verification specifications are prepared which cover each range of tests, inspections and demonstrations and define the acceptance criteria for design.

(c) *Under what circumstances would alternative calculations be performed?*

Where a characteristic cannot be verified other than by calculation, the calculations are either checked by someone else or the calculations are repeated using an alternative method.

(d) *Under what circumstances would design verification by similarity be valid?*

Where a design, a part of a design or a similar design has been previously proven to meet the same or more stringent requirements, verification by similarity may substitute for more costly verification methods. In such cases the objective evidence of previous tests or in-service use is produced.

(e) *When would tests and demonstrations be an appropriate design verification method?*

Design verification tests are conducted where a product possesses functional features to reveal design weaknesses and failure modes under anticipated operating conditions.

Demonstrations are arranged to verify non-testable product features.

(f) *How do you ensure that design stage documents are reviewed before release?*

All design documentation produced during a design phase is reviewed and approved before being released to the design review board for evaluation.

## Design validation

(a) *How is the design validated?*

Following successful design verification, validation tests are conducted to reveal design weaknesses and failure modes under operating conditions in excess of those anticipated in order to demonstrate a design margin adequate to compensate for the variability of material and process characteristics in production.

Following design validation a design certificate is issued by the design authority testifying the integrity of the design and qualifying it for particular applications.

When a design is required for an application for which it was not qualified, further validation tests may be performed providing the intended use is within the design limits.

## Design changes

(a) *How do you identify, document, review and approve all design changes and modifications?*

Following agreement to the design requirements all subsequent changes are subject to formal change control procedures. Following approval of a particular design solution at a design review, all subsequent changes are subject to formal change control procedures.

At each design review a design baseline is declared, comprising the identity of all requirements and specifications that have been approved. Any change to the design baseline is denoted by a unique reference and this identity applied to the product or service that embodies the change.

All proposed changes to an agreed design baseline are documented and passed to the Configuration Management Office, registered and subject to review by a design change control board chaired by the design authority. Prior permission to modify product in order to trial a design change must be obtained from the change control board.

The change control board evaluates all proposed design changes in terms of their impact on performance, cost and timescales and determines whether any change to the approved design is to be authorized, the time or batch from when the change is to become effective and what form the changes take.

Unless otherwise directed by the change control board the design is changed, applying the same practices and standards that were imposed for the original design.

Design documentation is changed in accordance with the policies pertaining to document control, the design subjected to re-verification and the design certificate endorsed accordingly.

Any design changes that require a change to products or services in production or use are classified as modifications and the point of embodiment determined prior to making the change. Documented modification instructions are issued following approval for the design change which define the action to be taken on products in production and products in service. All such instructions are reviewed and approved by the change control board prior to release.

If design changes are required for safety or environmental reasons, the change control board determines the extent to which product recall is necessary and initiates the appropriate procedures.

## Document and data control

### Document and data control procedures

(a) *What procedures have been established for controlling documents and data that relate to the requirements of ISO 9001?*

The following procedures are employed to control documents and data:

    CP    Document development

    CP    Document change

    CP    External document control

(b) *How do you control documents and data that relate to the requirements of ISO 9001?*

Any document and data for which its development, approval, issue, revision, distribution, maintenance, use, storage, obsolescence or disposal needs to be controlled is defined in company procedures and the appropriate control measures specified. All company documents are assigned an owner who may be the author or the sponsor. The Owner will be responsible for controlling such documents.

(c) *How do you control documents of external origin?*

The acquisition, use and maintenance of external documents, such as national standards, codes of practices, statutes, regulations and user manuals, are governed by documented procedures.

## Document and data approval and issue

(a) *How do you ensure that documents which relate to ISO 9001 are reviewed and approved for adequacy by authorized personnel prior to issue?*

Measures are taken to ensure that users of quality system documents and any prescriptive document derived from such documents are provided with an opportunity to establish the adequacy of such documents prior to their issue.

The procedures governing the preparation of such documents specify the approval authorities from which approval must be received prior to use. Document approval is denoted by signature on the document itself.

(b) *How do you ensure that data which relates to ISO 9001 is reviewed and approved for adequacy by authorized personnel prior to issue?*

External data used to convey requirements from design input through to servicing are selected from approved sources. Data that is generated by the company as input data to processes is validated by a qualified person from the same function that generated the data prior to it being placed in the user domain.

(c) *How do you preclude the use of non-applicable documents?*

Where a task requires the use of specific documents these are specified in the work instructions. A means of identifying the issue status of documents is provided which enables users to determine the correct version to be used.

(d) *How do you ensure that the pertinent issues of appropriate documents are available at all locations where operations essential to the effective functioning of the quality system are performed?*

Copies of quality system documents and any prescriptive document derived from such documents are issued on a controlled distribution which ensures that the users are provided with the current version. Staff are provided with a means of determining the correct version of a document to use.

(e) *How do you ensure that invalid and/or obsolete documents are assured against inadvertent use?*

Whenever a document is revised or withdrawn from use, change instructions are issued which specify the action required to avoid inadvertent use of obsolete documents. Draft documents are identified as draft and not used for operational purposes.

(f) *What means are used to identify obsolete documents retained for legal and/or knowledge preservation purposes?*

Copies of obsolete documents that are retained are marked 'Obsolete'.

## Document and data changes

(a) *How do you ensure that changes to documents are reviewed and approved by the same functions or organizations that performed the original review?*

When changes are made to an approved internal document, the author of the change submits the new version to the same review panel that was constituted for the original review.

No changes are made to an approved document without the approval of the authorities designated in the associated procedure.

(b) *How do you ensure that changes to data are reviewed and approved by the same functions or organizations that gave the original approval?*

When changes are made to approved data, the author of the change submits the new version to the same review panel that was constituted for the original review.

(c) *How do you ensure that designated organizations have access to pertinent background information upon which to make their review and approval of changes to documents?*

Details of the proposed change, the reason for change and its likely effects are provided to the approval authorities.

(d) *How do you identify the nature of changes within documents or their attachments?*

Changes within documents are identified so as to alert users to the changes and where necessary the reasons for change given.

# Purchasing

## General

(a) *What procedures have been established and documented for ensuring purchased product conforms to specified requirements?*

The following procedures are employed for purchasing products and services that impact deliverable product:

- CP    Purchasing Proprietary Hardware and Software
- CP    Subcontracting
- CP    Purchasing Services

(b) *How do you ensure that purchased product conforms to specified requirements?*

Any product or service procured for use with or for incorporation into products and services supplied to customers is governed by a documented purchasing specification which is conveyed to the chosen supplier. On receipt of the products or services offered by the supplier they are subject to verification against this specification before being accepted by the company. Only conforming products or services are subsequently used.

## Assessment of subcontractors

(a) *How do you evaluate and select your subcontractors?*

Suppliers to the company are classified as those supplying goods and services to their own specification, hereafter called 'vendors' and those supplying goods and services to a company specification, hereafter called 'subcontractors'.

Vendors are selected on the basis of past performance, recommendation or market standing. Subcontractors are selected on results of competitive tender and capability assessment by the company. To facilitate the selection of suitable suppliers, a list of preferred vendors and subcontractors is established and maintained.

Tenders are evaluated against financial stability, technical capability, management ability, adequate resources and quality system and quality assurance requirements.

(b) *In what document is the control exercised over subcontractors defined?*

The controls to be exercised over the subcontractor are based on demonstrated confidence in the subcontractor's ability to meet the prescribed requirements. Criteria are established and applied to determine the extent of any source surveillance, on site acceptance, receipt inspection or direct shipment for use.

Performance data on subcontractor's performance is collected and analysed, and the effectiveness of the subcontractor's controls determined. Corrective action is taken whenever the subcontractor's performance drops below a prescribed level.

(c) *In what documents do you record those subcontractors that are acceptable?*

The performance of current subcontractors is continuously monitored, the results recorded and the list of preferred subcontractors amended accordingly. The list denotes the products and services for which the subcontractor has been assessed, the standards against which the assessment was performed and any limitations on the use of a particular subcontractor.

## Purchasing data

(a) *How do you ensure that purchasing documents clearly describe the product ordered?*

Precise details of any products and services to be purchased by the company are given in writing to the Purchasing Manager who places all purchase orders on the company's suppliers. Purchase orders include details of the product or service, the governing specification, delivery date, the quality system requirements when appropriate and the company's standard conditions of purchase.

(b) *How do you ensure that purchasing documents are reviewed and approved for adequacy of specified requirements prior to release?*

All purchase orders are approved by the Purchasing Manager prior to release to the supplier. Purchase orders on subcontractors are reviewed by the design, manufacturing and quality assurance authorities for adequacy of the specified requirements prior to receiving the Purchasing Manager's approval.

## Verification of purchased product

### Supplier verification at subcontractor's premises

(a) *How are your subcontractor verification requirements and method of product release conveyed to subcontractors?*

To obtain an adequate level of assurance the quality assurance authority will determine whether it is necessary to carry out surveillance and/or product acceptance at subcontractor's premises and specify the appropriate quality assurance requirements in the subcontract.

### Customer verification of subcontracted product

(a) *How do you enable customers to verify purchased product at source or upon receipt?*

Where the customer has indicated in the contract a requirement for customer verification of purchased product or service at source or on receipt, the purchase order on the company's supplier conveys these requirements to the supplier. Any request by the customer to visit suppliers or verify goods on receipt is dealt with by the quality assurance authority.

## Control of customer supplied product

(a) *What procedures have been established and documented for controlling customer supplied product?*

Customers supply products for incorporation into finished product and also return company supplied product they have purchased for repair and maintenance.

The following procedures contain provisions for controlling customer supplied product:

| | |
|---|---|
| CP | Project Management |
| OP | Receiving |
| CP | Manufacturing |
| CP | Storage Control |
| OP | Dispatch |
| CP | Servicing |

(b) *How do you verify customer supplied product?*

Any product supplied by the customer either for incorporation into end product or for use during development or manufacture thereof is verified on receipt as meeting its description and is free from damage. No tests of customer property are carried out without authorization from the customer, provision of the relevant test instructions and relevant equipment.

(c) *How do you store customer supplied product?*

A register of all customer supplied products is established and maintained and the products stored separately to other items. Whilst in storage, customer property is stored as directed by the customer.

(d) *How do you maintain customer supplied product?*

No maintenance is carried out on customer property without authorization and provision of the relevant maintenance instructions and equipment.

(e) *How do you ensure that any lost or unsuitable customer supplied product is recorded and reported to the customer?*

In the event of customer property being lost, damaged or otherwise unsuitable for use, full details of the incident are recorded and reported to the customer by the quality assurance authority prior to any remedial action being taken. Any remedial action taken will be as directed by the customer.

## Product identification and traceability

(a) *What procedures have been established and documented for identifying product?*

The following procedures contain provisions for identifying product:

    CP    Hardware Design
    ST    Product Marking
    CP    Software Design
    CP    Purchasing
    CP    Manufacturing

(b) *How do you enable products to be identified from receipt and during all stages of production, delivery and installation?*

Products are given an identity at or before the first verification stage and retain this identity until incorporated into other product, following which their identities are traceable through the product specification.

Where the product is too small to carry the full identification, a coding convention is employed.

(c) *What procedures have been established and documented for the unique identification of individual product or batches?*

Products having functional characteristics carry a serial number to enable records to be traced to the product that has been processed, inspected or tested.

All items received into the company are registered and given a unique identity which is transferred onto records and which enables traceability from the end product back to the original purchase order.

All product records denote the inspection status of the item and the unit responsible for its supply and verification.

(d) *In what documents is the identity of individual product and batches recorded?*

Individual product identity is recorded on batch history records.

## Process control

(a) *How do you ensure that production processes that directly affect quality are identified, planned and carried out under controlled conditions?*

Following certification of the design a production plan is prepared which defines the activities, tools, equipment and processes required to produce the product and each of its components in the quantities required by the marketing plan or contract.

A manufacturing plan is prepared for each component to be manufactured (including assemblies) which defines the sequence of all operations to be performed to convert raw material or piece parts into finished product and to control the processes through which it passes.

(b) *How do you ensure that installation processes that directly affect quality are identified, planned and carried out under controlled conditions?*

Where the product is to be installed at its operational location, an installation plan is prepared which defines the activities, tools, equipment and processes required to install the product in the quantities required by the marketing plan or contract.

(c) *How do you ensure that servicing processes that directly affect quality are identified, planned and carried out under controlled conditions?*

Where product is to be serviced, a servicing schedule defines the levels, depth and frequency of preventive maintenance tasks to be performed. Servicing instructions are prepared detailing how specific maintenance operations are to be performed and the tests to be carried out to verify product serviceability.

(d) *What procedures have been established and documented which define the manner of production, installation and servicing?*

Those production, installation and servicing operations which do not rely on operator skill to achieve consistent results are defined in process specifications, installation procedures, servicing or maintenance procedures as appropriate.

Where tools, machines and equipment are capable of incorrect operation, operating instructions are provided which define the actions and precautions to be taken to ensure correct and safe operation.

(e) *In which documents do you define the production, installation and servicing equipment?*

Production, installation and servicing equipment is assessed to ensure that it is capable of producing conforming product consistently under the stipulated environmental conditions. The equipment required is specified in the appropriate manufacturing plan, installation plan, servicing plan or process specification.

(f) *In which documents do you define the production, installation and servicing environment?*

The working environment is specified in the appropriate manufacturing, installation or servicing plan or process specification when conditions other than normal atmospheric conditions are required for cleanliness, stability or other constraints.

Where the working environment needs to be controlled, measures are taken to train staff, prevent unauthorized access, alert staff to malfunction in the controls, record environmental conditions and regulate the ingress of matter that will be detrimental to the environment.

(g) *In which documents do you define the reference standards, codes of practice and quality plans to be complied with during production, installation and servicing?*

The standards, procedures and guidelines to be followed during production, installation and servicing are specified in the appropriate manufacturing, installation and servicing plans.

(h) *How do you monitor and control process and product characteristics during production, installation and servicing?*

Process specifications are prepared for processes requiring in-process monitoring to ensure consistent results. These specifications define the characteristics to be monitored, the instrumentation to be installed, the standard operating conditions, the records to be maintained and the criteria for starting and stopping the process.

(i) *How do you approve processes and equipment for use?*

Prior to launching into full production, a pre-production phase is completed in which the production tooling, equipment and processes are produced, commissioned, evaluated and proven capable of producing and maintaining the product characteristics required. During this phase the producibility of the product is evaluated in accordance with the approved product specification and changes initiated where alternative standards, methods of assembly or other aspects would improve the efficiency and effectiveness of production, installation and servicing.

Whenever process plant is modified, process capability is reassessed and following any maintenance, acceptance tests are performed to confirm conformance with standard operating conditions.

(j) *How do you define criteria for workmanship?*

Where the output of processes cannot be measured using available instruments, acceptance criteria are specified in the appropriate process specifications in terms of graphic images, samples, models or other suitable means. The documentation, identity, validity and status of any physical criteria are controlled by the quality assurance authority.

(k) *How do you ensure continued process capability?*

Process plant is subject to preventive and corrective maintenance to retain it in an operational condition and ensure continued process capability.

Whenever process plant is modified, process capability is reassessed and following any maintenance, acceptance tests are performed to confirm conformance with standard operating conditions.

(l) *In which documents do you identify the processes which produce results that cannot be fully verified by subsequent inspection and testing of product?*

All processes which produce results that cannot be verified by subsequent verification of the product are governed by process specifications controlled by the manufacturing authority.

(m) *How do you ensure that the results of these processes achieve the specified requirements?*

These process specifications define the conditions which need to be maintained, the procedures to be followed and the monitoring required to ensure that the specified requirements are achieved.

(n) *How are the requirements for process qualification specified?*

The requirements for process qualification including associated equipment and personnel are defined in the process specification.

(o) *In what documents do you record those processes, personnel and equipment that have been qualified?*

Records of qualified processes, equipment and facilities used in such processes are established and maintained by the manufacturing authority. Certificates are awarded to personnel who qualify for operation of certain processes and line managers are provided with maintained records of personnel qualified for particular processes.

## Inspection and testing

### General

(a) *What procedures have been established and documented for inspection and testing activities?*

The following procedures are employed for inspection and testing:

| | |
|---|---|
| OP | Receiving Inspection |
| OP | Machine Shop Inspection |
| OP | Assembly Inspection |
| OP | Electronic Component Test |
| OP | Printed Circuit Board Testing |
| OP | Final Product Testing |

(b) *How do you verify that the specified requirements for the product are met?*

Inspections and tests of the product are planned to take place at the point in the process where the characteristics being achieved are accessible for measurement. All inspections and tests ensure that no product is used, processed or released to the customer that does not meet the requirements specified for the product.

(c) *In which documents do you detail the required inspection and tests?*

The inspections and tests required are specified in a quality plan or in documented procedures. The acceptance criteria are specified in drawings or inspection and test specifications.

(d) *In which documents do you detail the inspection and test records to be established?*

The inspection and test procedures, specifications or quality plan define the information to be recorded and the form in which it is to be recorded when carrying out inspection and test activities.

### Receiving inspection and testing

(a) *How do you ensure that incoming product is not used or processed until verified as conforming with specified requirements?*

Prior to use or processing, all products and services procured for use with or for incorporation into products and services supplied to customers are registered, verified and identified as conforming with the requirements specified in the relevant purchase order by the procurement authority.

(b) *How is the amount and nature of receipt inspection determined?*

The extent of inspection needed on receipt of product is determined by the quality assurance authority and documented instructions are provided to the procurement authority.

Products accepted at source or destined for direct shipment to the production line bear an identity traceable to records previously deposited with the procurement authority to permit smooth transfer through Goods Inwards.

(c) *When you need to release incoming product for urgent processing, how do you enable immediate recall and replacement in the event of nonconformances being revealed?*

Products having characteristics that will not be verified at a stage subsequent to receipt inspection may not by-pass receipt inspection. Authority to release product prior to completion of receipt inspection is obtained from the quality assurance authority and the product positively identified and records of its location maintained in a manner that will enable its immediate recall should the need arise.

## In-process inspection and testing

(a) *How do you ensure that product is inspected, tested and identified as required by the quality plan or documented procedures?*

All in-process and final inspections and tests required to verify that the product conforms with the product specification is planned in conjunction with planning for manufacture and included in the manufacturing plan. With due regard to economical considerations, inspection and test stages will be planned to occur after a specified feature has been produced and before the feature becomes inaccessible to measurement.

Inspections and tests are conducted as specified in the manufacturing plan against acceptance criteria specified in the relevant product specification, supplemented where appropriate by inspection and test procedures defining the method of verification.

All products are identified prior to inspection and test as specified in the product specification. Following inspection and test, the verification status of the product is capable of being determined either by direct observation or through records traceable to the product.

(b) *How do you ensure product is held until the required inspections and tests or necessary reports have been received and verified?*

Where sampling takes place, the batch from which samples have been taken may continue through the process up to the next sampling inspection where it is held pending the results of the sampling inspection or test. Held product is segregated from conforming and from un-inspected product and carries an identification traceable to the sample.

## Final inspection and testing

(a) *In which document do you define the inspections and tests required to complete the evidence of conformance of the finished product to specified requirements?*

The inspections and tests required at the final stage and the records required to demonstrate conformance are specified either in the manufacturing plan or in a separate acceptance test plan and the method of inspection and test specified in inspection and test procedures.

(b) *In what documents do you require all the specified inspections and tests to have been carried out and require the results to meet specified requirements at the final inspection stage?*

Documented procedures require all products to be subjected to a final inspection prior to release to stores, delivery to other sites and prior to ownership passing to the customer. The final inspection determines that all planned manufacturing and verification operations have been completed and that the product conforms with the product specification.

(c) *How do you ensure that product is not dispatched until all the specified activities have been satisfactorily completed and the associated data and documentation is available and authorized?*

No product is released without the prior approval of the manufacturing authority and complete documentary evidence of conformance with the product specification.

## Inspection and test records

(a) *Which documents record the evidence that product has passed inspection and test with defined acceptance criteria?*

Records of all product or service verification activities are established and maintained throughout production and installation which demonstrate that the product or service conforms with the requirements of the product specification.

(b) *What measures are taken when a product fails to pass any inspection and/or test?*

Failed product is identified as nonconforming, placed in quarantine and subject to the nonconforming product procedures.

(c) *In which documents do you identify the inspection authority responsible for the release of conforming product?*

No product is despatched without the prior authorisation of the release authority endorsed on a release certificate. A register of stamp holders and approved signatories is established and maintained by the quality assurance authority for all personnel responsible for the release of conforming product.

## Inspection, measuring and test equipment

### General

(a) *What procedures have been established and documented to control, calibrate and maintain inspection, measuring and test equipment including test software?*

The following procedures are employed to control measuring devices:

| | |
|---|---|
| CP | Plant and Equipment Maintenance |
| CP | Electronic Equipment Calibration |
| CP | Mechanical Equipment Calibration |
| CP | Test Software Development |
| CP | Proprietary Software Maintenance |
| CP | Jigs, Tools and Fixtures |

(b) *How do you control devices used to demonstrate conformance of product with specified requirements?*

All devices used to demonstrate conformance of product with the product specification are registered with the quality assurance authority and their location, custodian and status recorded.

(c) *How do you calibrate devices used to demonstrate conformance of product with specified requirements?*

A means is provided to distinguish between devices which require periodic calibration and those devices that do not. Devices used for product acceptance purposes where the accuracy is susceptible to change are subject to periodic calibration. All such devices are calibrated prior to first use and the calibration intervals varied according to the nature of the device, the conditions of use and the seriousness of the consequences of any incorrect results.

(d) *How do you maintain devices used to demonstrate conformance of product with specified requirements?*

The user of a measurement device ensures that it is capable of the accuracy required prior to use and if damaged submits the device to the quality assurance authority for disposition. Preventive and corrective maintenance of devices used for product acceptance is undertaken as directed by the quality assurance authority.

(e) *How do you ensure that measurement uncertainty is known and consistent with the required measurement capability?*

Test and inspection procedures specify the measurements to be made taking into account any measurement uncertainty attributable to the device being used, the person carrying out the measurements and the environment in which the measurements are to be performed.

(f) *How do you ensure that devices used as suitable forms of inspection are proven capable of verifying the acceptability of product prior to their release for use during production and installation?*

All non-proprietary devices used for product acceptance purposes are governed by a product specification and subject to design verification prior to their release for use in production and installation.

(g) *What periodic checks are performed on such devices to ensure they remain capable of verifying the acceptability of product?*

A register of all non-proprietary devices used for product acceptance is established and maintained by the quality assurance authority and the devices subject to periodic re-verification.

(h) *What records are maintained of the checks carried out on devices used as suitable forms of inspection?*

The records of checks performed on devices that drift over time or with use form part of the device schedule.

(i) *In which documents do you define your measurement design data?*

Where original solutions to the measurement of characteristics are devised, the theory and evidence of the validity of measurement is documented and samples used for proving the method retained.

## Control procedure

(a) *How do you determine the measurements to be made and the accuracy required to demonstrate conformance of product to specified requirements?*

The product specification or any associated verification or process specifications identifies the measurements to be made and the accuracy required.

(b) *How do you ensure that appropriate inspection, measuring and test equipment is selected?*

Other than general-purpose devices, the particular measuring devices to be used are specified in the appropriate inspection and test procedures.

(c) *How do you ensure that inspection, measuring and test equipment is capable of the accuracy and precision necessary?*

All measuring devices are certified prior to first use and their accuracy and precision recorded on a calibration certificate. Accuracy and precision is an order of magnitude greater than necessary for any particular measurement so that users may rely on indicated values.

(d) *How do you ensure that all inspection, measuring and test equipment that can affect product quality is identified?*

All devices used for product acceptance purposes carry an identification label denoting its registration number, use classification and calibration due date where applicable.

(e) *How do you ensure that all inspection, measuring and test equipment that can affect product quality is calibrated at prescribed intervals or prior to use?*

A means of recalling devices for calibration is employed which minimizes disruption to work schedules. Calibration periods may be extended temporarily on the basis of calibration history and only with the prior written approval of the quality assurance authority.

(f) *How do you ensure that devices are adjusted when they have drifted outside the specified limits?*

All adjustments to adjustable devices are made by the laboratory appointed to perform calibration.

(g) *How do you ensure all calibrated inspection, measuring and test equipment has a known valid relationship to nationally recognized standards?*

Calibration of measuring devices is carried out by an accredited laboratory using measurement standards traceable to national standards.

(h) *In what documents do you record the basis used for calibration where no nationally recognized standards exist?*

In situations where there is no national standard against which to calibrate measurements, an internal standard is documented, its stability proven and approved by recognized experts.

(i) *In what documents is the process of calibration defined?*

On receipt of new measuring devices, the handbooks supplied are placed under the care and maintenance of the quality assurance authority and all calibrations carried out in accordance with the manufacturer's instructions. A data sheet is created and maintained for each measuring device which defines the calibration method, frequency and action to be taken should results be unsatisfactory.

(j) *What action is taken with devices when calibration results are unsatisfactory?*

Adjustment to measurement devices are made only by the calibration authority unless otherwise stated in the calibration data sheet. Adjustments are made only when devices are found to be outside the specified limits. Where no adjustment is possible, the device may be repaired subject to time and cost constraints and the approval of the budget holder. All repairs are carried out in accordance with the manufacturer's instructions.

(k) *What means do you use to identify the calibration status of inspection, measuring and test equipment?*

Devices subject to calibration carry a label denoting the calibration authority, the calibration due date and the device registration number.

(l) *What documents record the calibration of inspection, measuring and test equipment?*

Records are established and maintained by the calibration authority which enable the calibration and maintenance history to be determined. Records of results showing readings before and after calibration are maintained.

(m) *What action is taken with previous verification results when devices are found to be out of calibration?*

In the event of devices being found unserviceable during calibration, those using the device since the previous calibration will be informed and an assessment made as to the action required. Any product recall is made only with the prior approval of the Managing Director.

(n) *How do you ensure that environmental conditions are suitable for the calibrations, inspections, measurements and tests being carried out?*

Product specifications and associated verification and process specifications, where appropriate, define the environmental conditions required when carrying out product acceptance inspections and tests. Calibration of measuring devices is carried out under environmental conditions stipulated in the manufacturer's handbooks.

Areas requiring a controlled environment are equipped with a means of regulating and recording the environment.

(o) *How do you maintain the accuracy of inspection, measuring and test equipment during handling and storage?*

When not in use, measuring devices are stored in the containers provided by the manufacturer and the prescribed handling instructions adhered to.

(p) *How do you safeguard inspection, measuring and test equipment, test facilities and test software from adjustments which would invalidate the calibration setting?*

All devices subject to calibration are fitted with tamper-proof seals to prevent access to adjustable features.

Calibration facilities are secure from unauthorized access and any test software is accessible for read-only purposes following its release for operational use.

## Inspection and test status

(a) *How do you identify product in a way that indicates its conformance or nonconformance with regard to inspections and tests performed?*

Any product not bearing evidence of inspection or test is deemed to be uninspected. All products subjected to inspection or test bear evidence of whether or not they conform with specified requirements. When inspection status labels are to be applied to the product, the nature and location of such marking is specified in the product specification. In all other cases, records are maintained which enable the inspection status of items to be determined.

(b) *How do you ensure the identification of inspection and test status is maintained throughout production and installation?*

Once inspection status has been determined it is maintained through labels or related records so as to distinguish between conforming and nonconforming product.

(c) *What procedures are employed to ensure that only product that has passed the required inspections and tests is dispatched, used or installed?*

Dispatch procedures provide for checks to be carried out prior to packing to verify that only product bearing the correct acceptance identity is shipped.

## Control of nonconforming product

### General

(a) *What procedures have been established and documented controlling nonconforming product?*

The following procedures are employed to deal with product that fail to meet the requirements of the specification at an acceptance stage:

    CP    Nonconforming Product

    CP    Quarantine

(b) *How do you ensure that product which does not conform to specified requirements is prevented from inadvertent use or installation?*

All products are subject to the planned inspections and tests and those found nonconforming with the requirements are identified with a reject label and retained under the control of the inspection authority.

(c) *How do you ensure nonconforming product is identified?*

Details of the product, the nonconformance and the inspection authority are entered onto the reject label so as to reunite product with the label in the event of it becoming detached.

(d) *How do you ensure nonconforming product is documented?*

Details of the nonconformance are documented in a nonconformance report with provision for documenting the cause of the nonconformance and the agreed disposition.

(e) *How do you ensure nonconforming product is evaluated?*

Nonconformances are classified according to their severity and ease with which they can be eliminated. The inspection authority evaluates the nonconformance and determines whether its cause is common or special and assigns all special cause nonconformances to a review board comprising the design, quality assurance and manufacturing authorities.

(f) *How do you ensure nonconforming product is segregated?*

Where mixing of conforming and nonconforming product is possible, nonconforming product is placed in a quarantine area awaiting disposition. A register of all items in the quarantine area is maintained.

(g) *How do you ensure nonconforming product is dispositioned and concerned functions notified?*

At the earliest opportunity, the nonconformance is reported to the organization that supplied the product and disposition action requested.

## Nonconformity review and disposition

(a) *In which documents do you define the responsibility for the review and the authority for the disposition of nonconforming product?*

Where the product can be made to conform by rework or completing the planned operations, the inspection authority may return the product for such action and complete the nonconformance report accordingly.

The decision to scrap or repair product in-line with established repair procedures is made by the organization supplying the product. Should the product require redesign or non-standard repair, its disposition is made by the design authority.

Re-grading and use-as-is decisions may be made only by a review board comprising the design, quality assurance and manufacturing authorities.

(b) *How do you seek permission for the customer to supply or repair product that does not conform to specified requirements?*

Where the product cannot be made to conform with the customer's specified requirements but is considered fit for use by the review board and uneconomic to replace, a request for concession may be made to the customer to supply the product in nonconforming condition.

Should a repair still result in the product not conforming in full with customer requirements, a request for concession may be made to the customer to supply the product in nonconforming condition.

(c) *In which document do you record the description of nonconformity or repairs that has been accepted?*

Details of the actual condition of the accepted product are documented in the nonconformance report.

(d) *How do you ensure that repaired and reworked product is re-inspected in accordance with documented procedures?*

Following implementation, the product is returned for re-inspection or test either against the original requirement or as specified in the disposition statement.

## Corrective and preventive action

### General

(a) *What procedures have been established and documented for implementing corrective and preventive actions?*

The following procedures are employed to undertake corrective actions:

- CP    Quality System Maintenance
- CP    Configuration Management
- CP    Quality System Audit
- CP    Customer Feedback
- CP    External Audit
- CP    Supplier Assessment

The following procedures are employed to undertake preventive actions:

- CP    Quality Planning
- CP    Development Planning
- CP    Quality System Improvement
- CP    Human Resource Maintenance
- CP    Business Management

(b) *How do you ensure that the corrective or preventive actions taken is to the degree appropriate to the magnitude of the problems and commensurate to the risks encountered?*

Prior to taking a corrective action which requires additional resources, the cause is validated, a risk analysis carried out and management provided with sufficient data upon which to take the decision.

(c) *How do you implement and record changes resulting from corrective and preventive actions?*

All corrective action plans include any action necessary to change the affected quality system procedures and any derived or referenced documents. The resultant change is traceable to the initial corrective action report defining the source of the problem.

## Corrective action

(a) *How do you handle customer complaints and reports of nonconformities?*

All customer complaints are registered and remedial action taken to resolve the problem.

(b) *How do you ensure the cause of nonconformities relating to product, process and quality system is investigated?*

All procedures involving a quality verification stage provide for the results of the verification to be recorded and the records collected for subsequent analysis.

Data on nonconformances from whatever source is analysed by the quality assurance authority, grouped into either common cause or special cause categories and a corrective action report issued for action.

(c) *How do you eliminate the cause of nonconformities?*

Special cause nonconformances are dealt with promptly, the cause established and recurrence prevented.

Data on common-cause nonconformances is accumulated and analysed, the dominant cause established and measures taken to prevent recurrence.

(d) *How do you ensure that the corrective actions taken are effective?*

The corrective action to be taken, the responsibility for it and the date by which it is to be completed is recorded on the corrective action report. Provision is made for the action to be subsequently verified and its effectiveness evaluated and monitored to ensure that the desired goals have been achieved.

## Preventive action

(a) *What sources of information are used to detect, analyse and eliminate potential causes of nonconformities?*

Customer complaints, quality records, service reports and concessions are analysed by the quality assurance authority to search for opportunities for improvement and action initiated to prevent potential nonconformances.

Performance indicators are established for all result-producing processes and performance data collected, analysed and trends computed. Should the trend indicate that continuing the process will ultimately result in nonconforming product, timely corrective action is taken by the process owner.

Risk analysis is performed periodically to detect the failure modes that could impact system availability.

(b) *How are identified problems requiring preventive action dealt with?*

The quality system improvement procedure provides for any potential problems identified to be recorded on a problem report, a root cause analysis carried out and an improvement plan generated and implemented.

(c) *How is preventive action initiated and what controls ensure its effectiveness?*

Preventive action is initiated through an implementation plan approved by the quality manager or executive management depending on the resources required. Following implementation, the Quality Manager verifies that the action has been effective.

(d) *How do you ensure that information on preventive action is submitted for management review?*

The status of all approved preventive action plans is continually monitored by the Quality Manager and reported to executive management at the quality system review.

## Handling, storage, packaging and delivery

### General

(a) *What procedures have been established and documented for handling, storage, packaging and delivery?*

The following general handling, storage and packaging procedures are established, documented and maintained for the types of product that the company uses or produces.

| | |
|---|---|
| OP | Handling of Electrostatic Devices |
| CP | Control of Limited Life Items |
| GD | Handling Chemicals |
| ST | General Packaging Standards |
| OP | Control of Storage Areas |
| CP | Inventory Control |
| CP | Product Delivery |

The need for special handling, storage, packaging, preservation and marking provisions is established during product design and production planning and appropriate standards, procedures and guidelines documented and issued to those concerned.

### Handling

(a) *How do you prevent damage or deterioration of product in handling?*

Products are designed with handling provisions as appropriate. During the pre-production phase, the handling provisions needed to prevent damage and deterioration of product throughout its production are designed, installed, commissioned and certified operational.

The manufacturing plans, process specifications and procedures contain precautionary notes for handling product when applicable.

### Storage

(a) *What areas have been designated as locations in which product is prevented from damage or deterioration?*

Storage areas are provided for unused product and product awaiting further processing which will retain the product in its original condition until required for use and prevent damage and loss.

Where space permits and identity can be maintained, product is stored in its original packaging until required for use.

A means of identifying products requiring special storage conditions is provided which ensures that such products are only stored under the specified conditions. Where special storage conditions are necessary, means are provided to alerting staff to any deterioration in such conditions.

(b) *How do you prevent damage or deterioration of product in storage?*

Products remain in their original packaging until required for use. Any products that need to be dispensed or removed from their original packaging are stored under conditions that provide the same degree of protection.

(c) *How do you authorize receipt into storage areas or stock rooms?*

Except for designated quarantine areas, only conforming items are passed into controlled stores. Items are only received into stock on presentation of satisfactory evidence of receipt inspection and goods received note. Stock records are maintained which register sufficient data to retrieve items from a particular store and trace items back to original purchase order.

(d) *How do you authorize dispatch from storage areas or stock rooms?*

Items may only be removed from stores on presentation of an authorized requisition.

(e) *How do you detect deterioration in the condition of product in stock?*

A cursory check of the condition of stock is carried out at each withdrawal and a full stock condition check performed at intervals stated in the stores procedures.

## Packaging

(a) *How do you control packing, packaging and marking processes including the materials used?*

Bought-out product remain in their original packaging until required for use. Packing specifications are prepared or selected for company products and component parts which will preserve achieved quality from the time of manufacture through to incorporation into product or use by the customer.

Special packaging, preservation and marking requirements are identified during the product design phase and are governed by the relevant design policies.

Markings are applied such that the identity of the product can be established at any level of packaging.

## Preservation

(a) *What methods are employed to preserve and segregate product when the product is under your control?*

Limited life items, hazardous items or items susceptible to damage or deterioration by whatever means carry positive identification giving the necessary precautions to be observed.

Means of preserving product features as they are produced are employed where necessary to prevent damage or deterioration which might occur during further processing, storage or transportation.

Items remain segregated until required for use and measures taken to avoid cross contamination wherever possible.

## Delivery

(a) *How do you ensure that the quality of product after final inspection and test is protected up to its destination?*

Product are packaged in a manner and with such materials that will prevent damage or deterioration during transportation to its destination and pending first use.

## Quality records

(a) *What procedures have been established and documented for controlling quality records?*

A Records Control procedure is employed to control all records that demonstrate achievement of customer requirements and the effective operation of the quality system.

(b) *In which documents do you identify the quality records?*

Quality system procedures identify all the quality records that need to be produced and maintained to demonstrate achievement of the required quality and effective operation of the quality system.

(c) *What records demonstrate conformance to specified requirements?*

- CR   Design Verification Records
- CR   Design Validation Records
- CR   Design Review Records

CR     Subcontract Quality Records

CR     Calibration Records

CR     Deviations and Waivers

CR     Modification Records

CR     In-process Inspection and Test Records

CR     Final Inspection and Test Records

CR     Certificates of Conformity

(d) *What records demonstrate the effective operation of the quality system?*

The following records demonstrate the effective operation of the quality system:

CR     Quality System Audit Records

CR     Quality System Review Records

CR     Performance Assessment Reports

(e) *How do you identify your quality records?*

All quality records carry a generic name and a unique reference number and such other information as will enable the record to be traced to the product or service it describes.

(f) *How do you collect your quality records?*

When complete, quality records are collected and routed to the quality assurance authority for analysis as specified in quality system procedures.

(g) *How do you index your quality records?*

Registers of quality records are established and maintained so that any that are missing can be detected and located.

(h) *How do you provide access to quality records?*

The database manager provides access to staff with the need to know.

(i) *How do you file your quality records?*

Quality records are filed together with the corresponding register in the company's filing system and the file location specified in the relevant quality system procedure.

(j) *How do you maintain your quality records?*

In-process records are promptly updated with current information and placed in the designated files when not in use. Once complete, no change may be made to certified records unless endorsed by the original certifying authority.

(k) *How do you dispose of your quality records?*

When records have served current purposes they are removed to a controlled archive where they remain until dispositions are received by the quality assurance authority.

(l) *How do you ensure that pertinent subcontractor quality records are maintained?*

The quality records that subcontractors must maintain and the conditions of access and retention are specified in the subcontract. Any subcontractor records supplied to the company are retained in separate files to aid retrieval.

(m) *How do you ensure that all quality records remain legible?*

Where records are likely to become soiled in use, provisions are made to afford the appropriate protection.

(n) *How do you ensure that quality records are readily retrievable?*

In order to ensure that records are retrievable, the permanent location of current quality records are specified in quality system procedures and registers maintained to indicate any temporary location. Records located in the archive are stored in such a way as to facilitate their retrieval.

(o) *How do you ensure that quality records are stored in facilities that provide a suitable environment that minimizes deterioration, damage and prevents loss?*

The original quality records are retained in the designated files located in an office environment and placed in the care of an appointed custodian. After a specified interval, the records are removed to an archive store equipped with appropriate security and environmental controls.

Records that would have a detrimental effect on the business if lost by fire theft or unauthorised removal are identified as critical records in the procedures and measures taken to maintain insurance copies in a secure remote location.

(p) *In which documents do you record the retention times of quality records?*

Quality records are retained either for the duration of the contract, the life of the product or until they serve no further useful purpose. Disposal instructions are issued by the quality assurance authority.

(q) *How are quality records made available for evaluation by the customer?*

In the event of the customer requiring access to quality records beyond the period when they would normally be destroyed, or outside normal office hours, special provisions are included in the contract and conveyed to those concerned for implementation.

## Internal quality audits

(a) *What procedures have been established and documented for planning and implementing internal quality audits?*

A Quality System Audit Procedure is employed to plan and implement internal quality audits.

(b) *How do you verify whether quality activities and related results comply with planned arrangements?*

Internal quality audits are carried out by or on behalf of the quality assurance authority to determine whether the approved policies and practices are being followed and are effective in achieving the company's quality objectives. The audits also extend to verifying that the documents derived from the approved policies and procedures are being effectively produced and implemented and that the results comply with the specified requirements.

An annual audit programme is established, documented, agreed and conveyed to the managers affected prior to commencing the first audit. The objective of each individual audit is defined and the audit planned to ensure that the objective is achieved.

All audits are conducted in accordance with documented procedures designed to ensure a consistent approach and uniform records.

(c) *How do you determine the effectiveness of the quality system?*

The effectiveness of the quality system is determined by employing different auditing techniques and by collecting and analysing performance data generated by the system.

(d) *How do you ensure that audits are scheduled on the basis of the status and importance of the activity to be audited?*

Audits are scheduled so as to provide sufficient confidence in the operations audited. Operations critical to the business are audited more frequently than non-critical operations. New or changed practices and organizations are subject to audit soon after the changes have come into effect to verify stability following the change.

(e) *How do you ensure that personnel conducting the audit are independent of the activity being audited?*

The quality manager selects trained auditors on the basis of their independence from the activities they will be assigned to audit.

(f) *How do you ensure that the results of the audits are documented and brought to the attention of the personnel having responsibility for the area being audited?*

An audit report is issued to the manager of the areas audited which details the results of the audit.

(g) *How do you ensure that management personnel responsible for the area audited takes timely corrective action on deficiencies found by the audit?*

On receipt of the audit report the manager concerned determines any corrective action required in conjunction with the auditor and agrees a completion date. The completion date is appropriate to the circumstances, thus ensuring that action is taken before further deficiencies are generated. In the event that the manager concerned is unable to propose a suitable corrective action, he/she seeks out another manager with the appropriate authority.

(h) *How do you carry out audit follow-up actions?*

At a pre-defined time after the audit, a follow-up audit is undertaken to confirm that the agreed corrective actions have been carried out and were effective in eliminating the detected deficiencies.

## Training

(a) *What procedures have been established and documented for identifying training needs?*

Training needs are identified through the following procedures:

    CP    Human Resources Maintenance

    CP    Resource Planning

(b) *How do you identify training needs?*

Any routine training and re-training necessary for a person to perform a particular job effectively is specified in job specifications, process specifications, maintenance procedures etc., as appropriate.

Special training required as a result of new technologies, contracts, markets, company-wide improvement programmes etc. is specified in the associated plans.

(c) *How do you ensure that training is provided for all personnel performing work affecting quality?*

An annual training plan is established and maintained by the training authority which specifies details of the training to be carried out in each department sufficient to permit effective implementation of the quality system. Managers ensure that the training plan is implemented within their areas of responsibility.

(d) *How do you ensure that personnel performing specific assigned tasks are qualified on the basis of appropriate education, training and/or experience?*

Personnel records are established which describe the current qualifications, experience and training of each member of staff. Job specifications are prepared and maintained for each trade and position in the company. The job descriptions specify the qualifications, experience and training requirements for particular jobs. Managers select staff for particular jobs who meet the requirements of the job specification.

(e) *In what documents do you record the training provided?*

Details of any training carried out is recorded on personal training records and, where issued, certificates provided to the individual. For those jobs requiring a particular level of competence, staff lists giving competency levels are made available to managers.

Training records are periodically reviewed by managers to identify gaps in training and recency of training and training or re-training scheduled.

## Servicing

(a) *What procedures have been established and documented for performing, verifying and reporting servicing activities?*

The following procedures are employed to control servicing activities:

| | |
|---|---|
| CP | Servicing |
| CP | Service Contracts |
| CP | Technical Support |
| OP | Product Repairs |
| OP | Software Upgrades |

(b) *How do you ensure that servicing is performed in a way that meets the specified requirements?*

The activities required to retain one of the company's products in a serviceable condition are specified in the user manual supplied to the customer.

These manuals define the operating instructions, service restoration instructions, maintenance instructions, spare parts schedules, fault diagnosis charts and any precautionary measures required for safe operation and maintenance.

Where the company has agreed to service its products either on customer premises or on return to the factory, a servicing plan is prepared which defines service levels, procedures, equipment, technical support and escalation arrangements.

(c) *How do you ensure that servicing is verified in a way that meets the specified requirements?*

Any product returned by a customer is subject to survey in order to determine whether the company is liable for replacement or repair before taking further action. Such liability will be void if the product has not been serviced as specified in the user manual supplied to the customer.

Where the company has a service agreement with its customers, the servicing authority verifies that servicing operations are being carried out in accordance with such agreements and that overall service levels are being achieved.

(d) *What documents are used to report that servicing meets specified requirements?*

Service reports are used to collect data on preventive maintenance, routine tests and to record remedial actions carried out to resolve problems found on product during servicing.

## Statistical techniques

### Identification of need

(a) *How do you identify statistical techniques required for verifying the acceptability of process capability?*

Statistical techniques employed to determine process capability are specified in the production plan and based on proven statistical theory. Staff using the techniques are trained and provided with the necessary charts, data and instrumentation to make correct decisions from the evidence they have acquired.

(b) *How do you identify statistical techniques required for verifying the acceptability of product characteristics?*

Statistical techniques may be used to accept product when large quantities of homogeneous product are involved and these are specified in the relevant inspection or test procedures.

Where product characteristics cannot be verified except over prolonged periods and using a large sample size, statistical techniques may be used to predict performance providing the techniques are based on valid authenticated data. The prediction methods are defined in documented procedures and staff trained in their use.

**Procedures**

(a) *What procedures have been established and documented for implementing and controlling the application of the statistical techniques identified?*

The following procedures contain provisions for controlling the application of statistical techniques:

| | |
|---|---|
| OP | Receiving Inspection |
| OP | Electronic Component Test |
| OP | Printed Circuit Board Test |
| OP | Reliability Prediction |
| OP | Failure Mode and Effects Analysis |
| OP | Maintainability Prediction |

# Glossary

**Activity.**  An activity is the smallest parcel of work.

**Approved.**  Approved means that it has been confirmed as meeting the requirements.

**Assessment.**  The act of determining the extent of compliance with requirements.

**Assurance.**  Evidence (verbal or written) that gives confidence that something will or will not happen or has or has not happened.

**Authorized.**  A permit to do something, use something which may not necessarily be approved.

**Authority.**  The right to take actions and make decisions.

**Behavioural system.**  The dynamics of the internal culture and personnel relationships within an organization which produces characteristic behaviour patterns in its members.

**BPR.**  Business Process Re-engineering.

**Business level.**  The upper managers who focus on marketing and innovation and respond to the voice of the customer.

**Business process.**  A series of operations which are an essential part of a business.

**Business Process Mapping.**  Illustrating business processes in diagrammatic form indicating the relationships, interfaces across organizational boundaries.

**Business process model.**  A diagram of the key processes that convert external inputs into outputs showing their interrelationships and channels along which product or information flows.

**Business Process Re-engineering.** The fundamental rethinking and radical redesign of business processes to achieve dramatic improvements in critical, contemporary measures of performance, such as cost, quality, service, and speed (Michael Hammer).

**Business processes.** The composite of all processes that define how an organization conducts its business.

**Certification body.** An organization that is authorized to certify organizations. The body may be accredited or non-accredited.

**Commitment.** An obligation a person or organization undertakes to fulfil: i.e. doing what you say you will do.

**Continual improvement.** A form of improvement that brings about beneficial change in stages, reaching a plateau at each stage. The time on the plateau is immaterial providing there is no decline in performance.

**Continuous improvement.** A form of improvement that does not reach a plateau.

**Control.** The act of preventing or regulating change in parameters, situations or conditions.

**Control methods.** Particular ways of providing control which do not constrain the sequence of steps in which the methods are carried out.

**Control procedure.** A procedure that controls product or information as it passes through an organization.

**Controlled conditions.** Arrangements which provide control over all factors which influence the result.

**Core values.** Those key standards and principles which condition behaviour and opinion.

**Corrective action.** Action planned or taken to stop something from recurring.

**Critical Success Factors.** The factors upon which the achievement of a specific objective, goal or mission depend.

**Cultural analysis.** An analysis of the values, norms and beliefs of an organization to help understand.

**Culture.** The values, beliefs and norms that permeate an organization and help shape the behaviour of its members. A set of shared attitudes, values, goals and practices that characterizes a company or corporation (*Merriam Webster's Collegiate Dictionary*).

**Data.** Information which is organized in a form suitable for manual or computer analysis.

**Define and document.** To state in written form, the precise meaning, nature or characteristics of something.

**Demonstrate.** To prove by reasoning, objective evidence, experiment or practical application.

**Design.** A process of originating a conceptual solution to a requirement and expressing it in a form from which a product may be produced or a service delivered.

**Design and development.** Design creates the conceptual solution and development transforms the solution into a fully working model.

**Design review.** A formal documented and systematic critical study of a design by people other than the designer.

**Documented procedures.** Procedures that are formally laid down in a reproducible medium such as paper or magnetic disk.

**Effectiveness of the system.** The extent to which the (quality) system fulfils its purpose.

**Enterprise level.** The owners of the enterprise who focus on fiscal performance and listen to the voice of the shareholders or stakeholders.

**Establish and maintain.** To set up an entity on a permanent basis and retain or restore it in a state in which it can fulfil its purpose or required function.

**Evaluation.** To ascertain the relative goodness, quality or usefulness of an entity with respect to a specific purpose.

**Exposition.** A document that sets forth meaning or intent.

**Failure Mode Effects Analysis.** A technique for identifying potential failure modes and assessing existing and planned provisions to detect, contain or eliminate the occurrence of failure.

**Functions.** In the organizational sense, a function is a special or major activity (often unique in the organization) which is needed in order for the organization to fulfil its purpose and mission. Examples of functions are design, procurement, personnel, manufacture, marketing, maintenance etc. Departments may perform one or more functions but a department is a component of the organization, not a function.

**Guide.** A document that serves to assist someone perform an activity or make a decision.

**Implement.** To carry out a directive.

**Improvement plan.** A plan designed to bring about beneficial improvement irrespective of the scale of improvement.

**Management representative.** The person management appoints to act on their behalf to manage the quality system. Their actual title is irrelevant.

**Master list.** An original list from which copies can be made.

**Mission.** An organization's paramount objective for its intermediate future (Peter Druker). Where we are going in the short term.

**Multidisciplinary team.** A team comprising representatives from various functions or departments in an organization, formed to execute a project on behalf of that organization.

**Norm.** A principle of right action binding upon the members of a group and serving to guide, control or regulate proper and acceptable behaviour (*Merriam Webster's Collegiate Dictionary*).

**Objective.** The result that is to be achieved, usually by a given time.

**Objective evidence.** Findings that can be substantiated by information which is factual and which can be verified.

**Obsolete documents.** Documents that are no longer required for operational use. They may be useful as historic documents.

**Operating procedure.** A procedure that describes how specific tasks are to be performed.

**Operations level.** The middle managers and supervisors who focus on processes and respond to the voice of the process.

**Organizational goals.** Where the organization desires to be in markets, in innovation, in social and environmental matters, in competition and in financial health.

**PDCA.** Plan, Do, Check, Act.

**Plan.** Provisions made to achieve an objective.

**Policy.** A guide to thinking, action and decision. Policies can exist at any level in an organization from corporate level to the lowest level where activities are performed.

**Prevent.** To stop something from occurring by a deliberate planned action.

**Preventive action.** Action proposed or taken to stop something from occurring.

**Procedure.** A sequence of steps to execute a routine activity. Procedures can address interdepartmental, departmental, process, group, section or individual activities.

**Process.** A sequence of tasks which combine the use of people, machines, methods, tools, environment, instrumentation and materials to convert given inputs into outputs of added value.

**Process capability.** The ability of a process to maintain product characteristics within pre-set limits.

**Process management.** The planning, organization and control of those processes in a business that contribute to the delivery of business outputs.

**Process parameters.** Those variables, boundaries or constants of a process that restrict or determine the results.

**Product.** Anything produced by human effort, natural or man-made processes.

**Project.** A planned undertaking to achieve a specified result by a given date.

**Project phases.** The distinct periods through which a project passes from its launch to completion. Each phase has a measurable objective with defined deliverables to judge completion.

**Quality objectives.** Those results that the organization needs to achieve in order to improve its ability to meet current and future customer needs and expectations.

**Quality problems.** The difference between the achieved quality and the required quality.

**Quality records.** Objective evidence of the achieved features and characteristics of a product or service and the processes applied to its development, design, production, installation, maintenance and disposal as well as records of assessments, audits and other examinations of an organization to determine its capability to achieve given quality requirements.

**Quality requirements.** Those requirements that pertain to the features and characteristics of a product or service which are required to be fulfilled in order to satisfy a given need.

**Quality system.** A tool for achieving, sustaining and improving quality. Such a system should integrate interconnected business processes which collectively cause the supply of conforming product/service and prevent the supply of nonconforming product/service.

**Quality system assessments.** External audits carried out by second or third parties.

**Quality system element.** A distinct part of the system which is governed by a set of requirements. A subsection of the standard identified by a two-digit number such as 4.1, 4.2, 4.3 etc.

**Quality system management.** A business process the purpose of which is to plan, organize and control the development, operation, maintenance and improvement of the quality system so that it enables the organization continually to meet its corporate objectives.

**Quality system requirements.** Requirements pertaining to the design, development, implementation and maintenance of quality systems.

**Quality target.** An intermediate goal to be achieved towards a specified quality objective.

**Registrar.** See *Certification body*.

**Remedial action.** Action proposed or taken to remove a nonconformity (see also *Corrective and preventive action*). The action applies to the affected item, process or activity.

**Responsibility.** An area in which one is entitled to act on one's own accord.

**Review.** Another look at something.

**Service.** Results which do not depend on the provision of products.

**Shall.** A provision that is binding.

**Should.** A provision that is optional.

**Subsystem.** Subsystems comprise a group of business processes that serve a common purpose in the organization. The subsystem group is only necessary in modelling the system when intermediate layers are needed before identifying work processes.

**Surveillance (quality system).** An activity performed to verify that the organization has maintained its quality system and it continues to be suitable for achieving its stated objectives and effective in providing an adequate degree of control over the organization's operations.

**System audit.** An audit carried out to establish whether the quality system conforms to a prescribed standard in both its design and its implementation.

**System Change Board.** A group of people formed to evaluate proposed changes to the quality system and decide an appropriate course of action.

**System effectiveness.** The ability of a system to achieve its stated purpose and objectives.

**System Review Board.** A group of people formed to review the quality system, make decisions on its effectiveness and set a course of action for its improvement.

**Task.** A task is a group of activities that accomplish a minor objective.

**Theory of constraints.** A thinking process optimizing system performance. It examines the system and focuses on the constraints that limit overall system performance. It looks for the weakest link in the chain of processes that produce organizational performance and seeks to eliminate it and optimize system performance.

**Total Quality Management.** A management approach of an organization centred on quality based on the participation of all of its members and aiming at long-term success through customer satisfaction, and benefits to all members of the organization and to society (ISO 8402). A comprehensive approach to improving competitiveness, effectiveness and flexibility through planning, organizing ad understanding each activity, and involving each individual at each level (John Oakland).

**TQM.** Total Quality Management.

**Values.** The standards and principles that determine our behaviour and our opinions.

**Vision.** Where we want to be in the long term.

**Work instructions.** Instructions that prescribe work to be executed, who is to do it, when it is to start and be complete and how, if necessary, it is to be carried out.

**Work process.** A sequence of tasks that combine the use of people, machines, tools, methods, environment, instrumentation and materials to convert given inputs into outputs of added value.

# Index

## A

Approval controls, 243
Assessment – organizing, 107
Assessment of subcontractors, 371
Assigning the auditor, 301
Audit – external, 309
Audit conclusions, 303
Audit conduct, 302
Audit data analysis, 339
Audit findings, 303
Audit follow-up, 305
Audit preparation, 302
Audit reporting, 303
Audit schedule, 300
Auditing tools, 305
Auditors – assigning, 301
Authors, 92

## B

Behavioural system, 31
Breakthrough and control, 22
BS 7799, 16
BS 8800, 16
Business changes, 327
Business process re-engineering, ix, xiii
Business processes, ix, 42, 46, 70, 71, 82, 118, 124, 127, 143, 181, 207, 211, 236, 285

## C

Causing success, 28
Chain of processes, 18, 30
Change data, 340
Changes in requirements, 328
Charting processes, 145
Classifying audit findings, 303
Commitment, 61, 101
Computer network, 102
Consultant – selecting, 64
Consultant support, 93
Context diagrams, 130
Contract review, 360
Control procedures, 165
Control sequence, 24
Controlling data, 259
Corrective action plans, 332, 334
Corrective action proposals, 315
Corrective action records, 334
Corrective action, 315, 387
Corrective action system, 328
Corrective action tools, 329
Critical success factors, 35, 49, 50, 85, 104, 118, 159
Cross-functional charts, 148
Cultural analysis, 53
Customer needs and expectations, 26
Customer-supplied product – control of, 372

# D

Data collection, 308, 339
Data requirements, 84
Database controls, 242, 260
Database design, 240
Database manager, 261
Database specification, 241
Database testing, 244
Delivery, 389, 391
Deming, Edwards, 37
Deploying product/service characteristics, 159
Deploying requirements, 151
Deployment – function, 141
Derived documents, 166
Design and development planning, 363
Design changes, 367
Design control, 362
Design input, 364
Design output, 364
Design reviews, 87, 103, 365
Design validation, 367
Design verification, 366
Development methodology, 128
Document accessibility, 257
Document and data approval and issue, 368
Document and data changes, 369
Document and data control, 239, 368
Document attributes, 241
Document change notice, 278
Document change procedures, 250, 271
Document change process, 272
Document change request, 277
Document coding, 200
Document control requirements, 244
Document control strategies, 239
Document development plan, 188
Document development procedures, 246, 266
Document development process, 267
Document identification, 178
Document requirements, 187
Document what you do, 5, 6, 9, 122
Documentation analysis, 170
Documentation development, 187
Documentation hierarchy, 177
Documentation methods, 189
Documentation needs, 162
Documentation standards, 193, 233

Documents – types of, 164
Documents requirement by ISO 9001, 184

# E

Education, 99
Electronic documents, 240
Eviromental management, 21, 16
Executive briefing, 4
Existing practices, 98
External audit, 309
External document control procedure, 279
External document control process, 280
External documents, 258, 368

# F

Failure prevention system, 27
Flow charting, 129
FMEA, 161
Follow-up audit, 316
Forms, 226
Function deployment, 141
Function representatives, 92

# G

Generic process, 144

# H

Handling, 389
Health and safety management, 16

# I

Immunity to change, 203
Inspection and test records, 379
Inspection and test status, 384
Inspection and testing, 377
 final, 379
 in-process, 378
 receiving, 378
Inspection procedures, 377
Inspection, measuring and test equipment, 380
Internal quality audits, 394

ISO 14000, 21,16
ISO 8402, 10, 42
ISO 9001 Exposition, 351
ISO 9002, 6, 7,
ISO 9003, 6, 19
ISO 9004, 176
ISO 9004, 6, 10, 11,

## J

Job descriptions, 225
Juran J. M., 35

## L

Layout of documents, 198
Leadership, 111

## M

Management representative, 344, 355
Management responsibility, 352
Management review, 316, 355
Management systems, 12, 15, 16,
Mistakes – common, 74, 109

## N

Network security, 261
Nonconforming product – control of, 385
Nonconformity report, 9, 385
Nonconformity review and disposition, 386

## O

Oakland, John, 39
Objective evidence, 25
Objectives – setting, 66
Obsolete documents, 257
Operating procedures, 165
Operational policies, 209
Organization changes, 327
Organization charts, 224
Organization structure, 35
Organizational and technical interfaces, 363
Organizing the project, 89

## P

Packaging, 389, 391
Packard, David, 35
PDCA, 37
Performance assessment, 307
Performance indicators, 307
Policy Manual, 206
Preliminary review, 52
Preservation, 389, 391
Preventive action, 387, 388
Priorities – setting, 83
Problem database, 107
Problem identification, 329
Problem report, 329, 339, 340
Procedures, 38
   control, 165
   creating, 212
   document change, 250, 271
   document development, 246, 266
   external document control, 279
   operating, 165
   purchasing, 370
   quality system, 357
Procedures and instructions, 213
Process control, 374
Process decomposition, 146
Process management, ix, 110, 286
Process modification, 154
Process owners, 91
   role description, 236
Processes, 37
   business management, 173
   document change, 272
   document development, 222, 267
   external audit, 311
   external document control, 280
   generic, 144
   interconnection, 10, 14
   performance assessment, 307
   product aquisition, 147, 155
   quality system management, 286
   resource management, 150
   system audit, 299
   system implementation, 287
   system improvement, 338
   system maintenance, 325
   system review, 317

Product and service quality functions, 347
Product identification and traceability, 373
Product/service characteristics – deploying, 159
Project feasibility, 49
Project management strategies, 81
Project manager
　appointing, 72
　role, 73
Project office, 95
Project organization, 89
Project plan, 87
Project planning, 45, 82
Project reviews, 89, 103
Proving the need, 47
Purchasing data, 371
Purchasing procedures, 370
Purchasing, 370

## Q

Quality audits – internal, 394
Quality director – role, 346
Quality function, 347
Quality improvement plan, 341
Quality information system, 96
Quality Manual, 357
　changes, 316
　endorsing, 316
　reviewing, 312
　submitting, 312
Quality planning, 358
Quality policy, 204, 352
Quality records, 391
Quality system
　defining purpose, 68
　scope, 67
Quality system composition, 120
Quality system definitions, 21
Quality system design, 126
Quality system development manager – role, 348
Quality system development process, 126
Quality system development programme, 88
Quality system evaluation manager – role, 348
Quality system management process, 285, 286
Quality system principles, 21

Quality system procedures, 357
Quality system requirements, 356
*Quality World*, 32

## R

Records, 227
　corrective action, 334
　inspection and test, 379
　quality, 391
　system review, 320
Reference documents, 168
Remedial and corrective action, 305
Resource changes, 327
Resource data, 339
Resources, 38, 58, 354
Responsibility and authority, 353
Reviewing documents, 243
Revision controls, 243
Risk analysis, 159
Role descriptions, 226
Role of process owner, 236
Role of Project Manager, 73
Role of quality director, 346
Role of quality system development manager, 348
Role of quality system evaluation manager, 348
Root cause analysis, 305, 331, 341

## S

Security management, 16
Servicing, 396
Site co-ordinator, 93
Small businesses, 17
Software – choice of, 241
Software tools, 157
Sole trader, 18
Staff changes, 328
Statement of work, 94
Statistical techniques, 397
Steering group, 89, 90
Storage, 389, 390
System audit, 297
System change board, 333
System co-ordinator, 93

System design strategy, 121
System design, 115
System evaluation, 296
System improvement, 337
System maintenance, 324
System management organization, 344
System models, 133, 136-141
System monitoring, 326
System requirements, 117
System review frequency, 320
System review meeting, 319
System review record, 320
System review report, 340
System review strategies, 316
System review, 316
Systems – nature of, 115
Systems engineering, ix, xi, xiii, 17, 20, 70, 333

## T

Task analysis, 151

Teams, 81, 89, 91, 95, 99
Technology changes, 328
Total Quality Management (TQM), 34
   relative position of systems, 34
   relationships to ISO 9000, 42
Training, 99, 395
Trend analysis, 339

## V

Verbal instructions, 12
Verification
   customer, 372
   supplier, 372
Verification of purchased product, 372
Verifying effectiveness, 334, 342

## W

Work process, 124, 127
Writing style, 195